SPARTANAT BLACK BOOK

1

SPARTANAT

spartanat.com

ISBN 978-3-903526-04-4

THE TACTICAL DRONE

MILITARY USE OF COMMERCIAL AIR VEHICLES

OBSERVE
RECCE
DOCUMENT
ATTACK

CHRISTIAN VÄTH &
MARKUS REISNER

TABLE OF CONTENTS

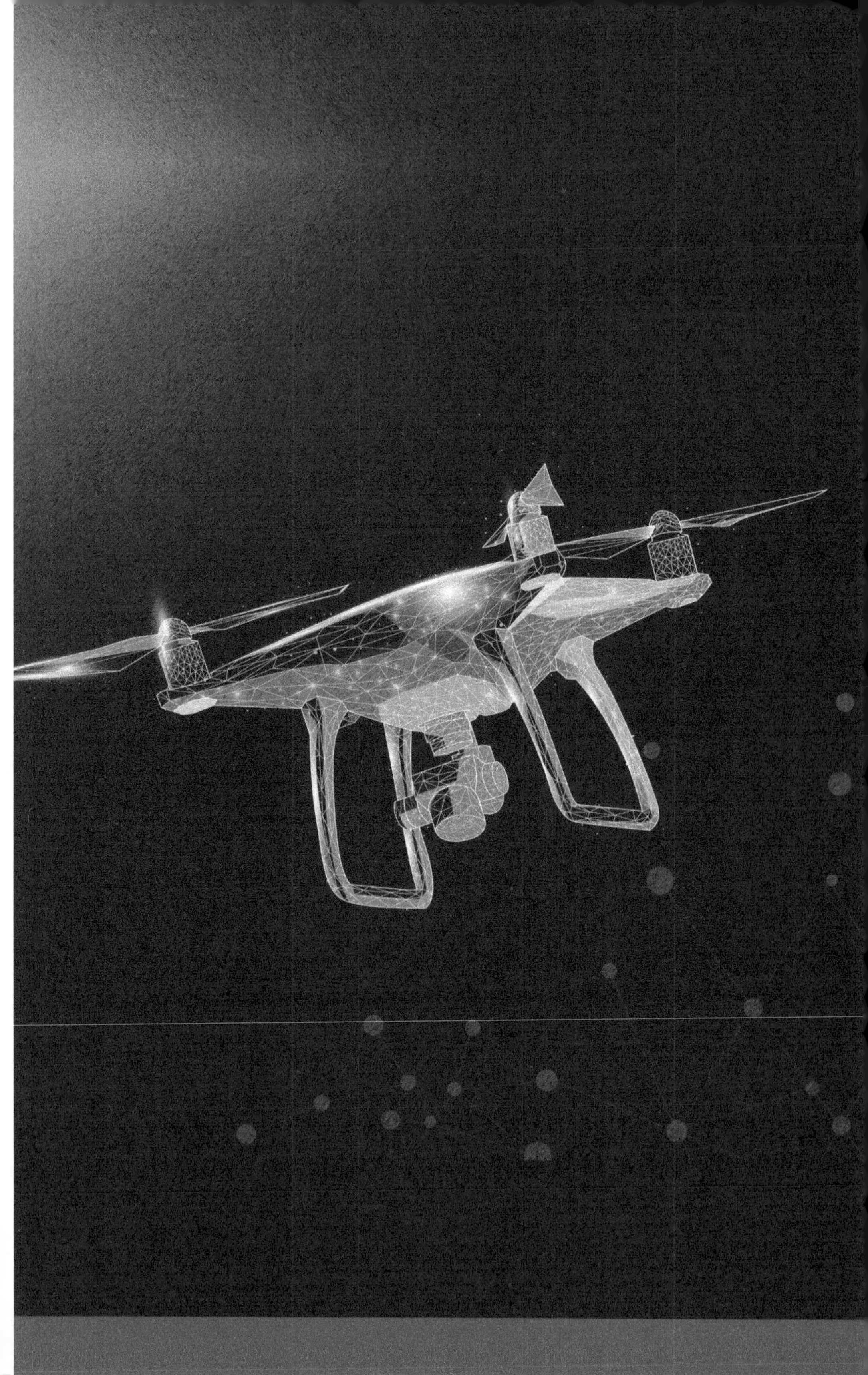

INTRODUCTION:

ALL GOOD THINGS COME FROM ABOVE

1

THE SKY HAS ALWAYS BEEN DANGEROUS FOR SOLDIERS AND OTHER COMBATANTS, EVER SINCE MAN INVENTED FLYING.

From above, the overview is infinitely better. Map-like pictures make observation, coordination, and attacks much easier. From the balloons that monitored what was happening at the front to the first cruise missile that flew against England, the "V1," there have always been many means other than manned aircraft.

Modern technology has led to a radical change and gives everyone the opportunity to operate affordable high-quality systems that can be used in many ways. The conflicts of the past 20 years have created a need for a large variety of unmanned systems with a wide range of specializations and sophistication, from model aircraft (like the Russian "Orlan-10") to special military drones, which are equipped with high-tech for reconnaissance. Particularly exciting and interesting is the democratization of this technology. Anyone can get a drone of greater or lesser value through Amazon or other stores and put it to work immediately.

The drone has become an important and versatile weapon in modern conflict. We have moved on from the large "Predators" that stalked their victims as silent observers and stealthy missile platforms in Obama's War on Terror. Terrorists in particular were the first to begin using commercial drones as tools and weapons across the world, most notably the "Islamic State," which produced the first videos of "DJI Phantom" drones observing and bombing in Syria. The war in Azerbaijan has shown how the massive use of drones, in this case professional rather than commercial systems, can completely overwhelm an opponent. The war in Ukraine may be called a breakthrough in the future, as the Ukrainian forces showed surprising efficiency in the use of commercial drones. The first large-scale proof of concept was the Battle of Kyiv in 2022. Since then, commercial drones have been integrated into every tactical operation on this battlefield—by both sides. They can be used as scouts, observers, or documentation devices to provide footage in the war for information superiority. They can even drop hand grenades on the enemy with amazing accuracy, making them poor men's bombers. In every way, small commercial drones can be called the new best friend of the infantry squad.

"The Tactical Drone" is the first handbook for the use of civilian drones in a professional environment. These little helpers are valuable companions for people who want to increase the possibilities of the available forces and also increase protection. These unmanned systems can provide valuable information in real time for the lowest ranks, making them a powerful tool for low-level commanders. Every commercial drone can become a tool that makes the force more efficient when used properly. In this book, we offer guidelines on equipment, careful handling, and application of the drone as a tactical tool. At the same time, it gives valid information for the legal use of small drones in a civilian environment.

Christian VÄTH

Markus REISNER

THE TACTICAL DRONE

HOW TO INTEGRATE SMALL RECONNAISSANCE DRONES INTO THE INFANTRY SQUAD'S METHODS OF COMBAT.

In principle, small drone systems can be carried, maintained, and deployed by a single operator. Since both the implementation of the reconnaissance mission and the interpretation of the results require at least a rudimentary tactical overview, it is a good idea to maintain close contact with the squad leader. A rifleman can act as a drone operator in his secondary role, just like the radio operator. Alternatively, a two- or three-man team consisting of an operator and a team leader is established. Similar to the distribution of tasks in a sniper team ("the shooter shoots, the observer hits"), one soldier flies the drone and the other one observes. An optional third man would be of use in securing the close surroundings during flight time. The following examples are intended to illustrate the possibilities and limitations of the usability of **sUAS** *(Small Unmanned Aerial Systems)* for the light infantry squad.

EXAMPLE 1:

SECURITY

After three exhausting days of combat, our squad A1 establishes a security perimeter on the outskirts of a small Central European town as a part of the 1st platoon. On the previous day, the squad suffered one KIA and two WIA. Besides the squad leader, the element is down to only six men available. The machine-gun team has established a position with a good field of fire across the adjacent main road toward the center of the town, where the bulk of the enemy forces are believed to regroup. On the left flank is a large parking lot and easily surveyed terrain. In the rear, a farm lies a short distance away; the platoon command post and the casualty collection point have been set up there. It also serves as a rest area for A1. The right flank consists of confusing small alleys and damaged buildings. A few houses further behind lie the marketplace and larger streets in all directions. Besides the squad leader, five men are currently tied up: The machine gun team observes its field of fire from the position addressed, a riflemen team secures the problematic right flank at least partially, and one man tries to organize rations and ammunition via the platoon. Only the squad leader and one rifleman are left. Shift capability cannot be achieved in this way, and the capability to fight will decline over time. Promised field replacements will not arrive for at least a week and will probably be of a lower level of training. How can sUAS now improve the situation of the squad and thus the platoon?

The squad leader suggests to the platoon leader that he monitor his right flank in depth through drone flights and passive security measures (light and noise traps). This would also relieve neighboring squad A2. A flight plan is drawn up, and the problem area is patrolled from the air at irregular and short intervals. The flights will be alternated by the drone operators of the two squads. The two riflemen deployed for security prepare light traps in a short time using wire and high-power chem lights. These are supplemented with noise traps using various materials from the surrounding houses. The squad can now move into shift operations, with three squads taking turns manning the positions and the squad leader taking over the initial drone flights himself. A1 can remain combat-ready without having to accept major gaps in security.

EXAMPLE 2:

ATTACK

Three days later, the battalion of our squad A1 has fought through about half of the small town. In the process, A1 experienced no further casualties. The Alpha platoon receives the order to move into an assault position for an attack on the enemy's left flank. To do this, A1 is to reconnoiter the main emphasis in the enemy's defenses using sUAS. The squad leader can hardly recon positions because they are mostly in the houses in the depths of dark rooms. The platoon leader suggests to the company commander that he provoke a reaction from the enemy at another place through massive fire and a smokescreen. The attack order is adjusted to allow this deception. A short time later, Charlie platoon opens fire with all weapons on previously identified or suspected enemy positions and then moves into alternate positions. Simultaneously, A1's drone operator team reconnoiters the enemy's movements between the buildings. The attempt is clear: The enemy tries to change its main emphasis to counter the expected attack in Charlie's sector. Having clearly observed this move, the company attacks on the left flank, where the enemy positions are now significantly weaker. The assault squads make rapid progress and have taken the first sections of buildings even before the previously displaced forces can return to their old positions. The enemy infantry can be encircled.

EXAMPLE 3:
PATROL

Another two days later, the small town is captured; the battalion is resupplied and can rest for at least one day. The squad leader of A1 receives a warning order to conduct a patrol on one of the three hills lying in the direction of the enemy, immediately beyond the edge of the town. The battalion commander suspects an encirclement of the entire area by mechanized forces of the enemy and the use of artillery observers on these hills. In addition to reconnoitering the remaining enemy forces in this hilly terrain, the reconnaissance mission also includes reconnoitering suitable ambush positions to slow down the expected attack of the enemy. The squad leader turns the squad over to his deputy with orders to rest, resupply, and then continue defensive preparations in accordance with his orders. He takes only one particularly capable rifleman with him and takes charge of the sUAS materiel himself. After a short coordination with the commanders of the farthest positions, the patrol finds itself in no man's land between the fronts. In short, covered sprints, the patrol works its way through previously visible terrain to the base of the hill. The slope is covered with fields. Hedges provide some screen, but there is no adequate cover. The squad leader prepares the equipment for a drone flight. He carefully plans the flight path: His sUAS is to stay close to the ground behind the hedges and move crosswise—as far away from his own position as possible. He then wants to fly in a large arc to the top of the hill. If the enemy is aware of the drone at this point, the exact position of the patrol will not be obvious to him due to the different axes of movement.

Shortly before the drone reaches the summit, it is shot down. The squad leader could not see where the fire came from. The patrol remains in its position and radios the battalion. The battalion commander immediately gives the order to fire the prepared mortar fire and covers the entire summit and the rear slope with 120 mm rounds. After the last impact, the patrol quickly jumps over the craters to the summit and conducts a battle damage assessment of the spot. This results in a firefight with a surviving enemy rifle team, which the patrol is able to win. One enemy soldier is able to flee downhill. The final assessment reveals that an artillery observation team and a rifle team were probably just setting up positions on the summit. Some equipment was still packed. The patrol completes its mission, hands the hill over to a forward security detachment, and returns to its squad.

EXAMPLE 4:

DEFENSE

The next day, a massive mechanized attack with artillery support begins. Overnight, in anticipation of such an attack, light infantry forces were deployed in ambush positions in all directions. In the narrow paths of the right flank, the intense and surprising fire of the infantry, followed by mortar fire, brought the advance to a halt. However, on the left flank, which consists of open fields, the enemy attack succeeded, and the enemy tanks are on the verge of advancing into the rear and cutting our battalion off.

It is suspected that the enemy will take the opportunity to use mechanized infantry to get to the edge of the village in the rear and then, dismounted, quickly take the first buildings to re-establish themselves in the village. As the tanks try to encircle the town and artillery rounds detonate across the city, A1 is near the battalion command post to gather supplies for the company. The battalion commander hastily attempts to reinforce the rear perimeter of the village to counter the expected attack. A1 quickly moves into new positions with good fields of fire. The drivers of the brigade are unceremoniously commandeered by the squad leader; every rifle counts. After a brief terrain assessment, the tank destruction team receives its warning order. But the attack does not come at the expected point. Instead, the enemy pushes through a gap in the defenses a few houses away. The squad leader realizes the seriousness of the situation and does not wait for his order. The now-reinforced squad runs through narrow but familiar alleys. Sudden enemy fire brings the movement to an abrupt halt. Bullets from machine guns slam into the surrounding buildings. "Hold position!" is the order. The squad leader runs back past two buildings into a courtyard and orders the deployment of the drone. The backup drone ascends and immediately reveals a hostile Infantry Fighting Vehicle about a hundred meters away, firing on the squad position. After a short arc flight, the position of the enemy assault group can also be roughly guessed. The group's second and last drone is shot down a moment later. It does not matter; the leader has seen enough. The tank destruction team is quickly reassigned and is led by the squad leader into a suitable firing position. Two shaped charge grenades hit the vehicle in quick succession, destroying it. A second Armored Personnel Carrier in the depth, shooting a smoke screen, drives backwards out of sight into a bump near the road. The opportunity is seized. The dismounted forces are separated from their vehicles. The squad leader quickly pushes the team back into squad position and catches the enemy in his flank. At the same time the first reinforcements arrive, the defense adapts to the changed situation. The rear is soon no longer a rear, the moment of surprise could not be exploited by the enemy.

The combat examples described here are intended to provide a reference to define technical requirements for sUAS as well as operational capabilities and limitations. Small reconnaissance drones at the squad and/or platoon level can:

- **SPARE OR EXPAND THE EMPLOYMENT OF OWN FORCES,**

- **REDUCE OWN CASUALTIES AND INCREASE ENEMY LOSSES,**

- **ACCELERATE OR ENHANCE DECISION-MAKING,**

- **IMPROVE SITUATIONAL AWARENESS AND ACHIEVE LOCAL INFORMATION SUPERIORITY,**

- **PROVIDE REACTION TIME FOR OWN TROOPS AND DECEIVE THE ENEMY,**

- **GENERATE AND DOCUMENT DETAILED ENEMY INTELLIGENCE (TYPE, STRENGTH, BEHAVIOR),**

- **GUIDE DIRECT AND INDIRECT FIRE OF ALL CALIBERS QUICKLY AND ACCURATELY.**

TECHNICAL REQUIREMENTS

The critical performance parameters of sUAS are primarily flying time, range, speed, and weight. Flying time determines the total mission duration before at least one interruption is required for battery replacement. Currently, 30 minutes is technically easy to ensure, which is sufficient for the vast majority of tactical applications at the level considered here. The same applies to range. Products available on the market already have a range of up to ten kilometers, which significantly exceeds the usual reconnaissance depth at the subunit level of infantry forces.

Due to its small size, even a small drone hovering on the spot at the appropriate altitude is a small and difficult target to hit for all available weapon systems. Thus, a higher flight speed primarily increases the time available over the reconnaissance target. Current designs are of low weight already, and the performance data described can already be achieved with a system weight of 0.55 pounds. Including the remote control, power supply, screen, and spare parts, the mission equipment remains portable and is easy to stow. This makes it possible to carry a spare drone. Furthermore, the capabilities of the installed reconnaissance systems can be decisive.

It is desirable to have the highest resolution imaging possible displayed without delay via stable data transmission on a compact and robust display. In addition, the reconnaissance capability in poor visibility or darkness, for example through thermal imaging technology, represents a fundamental gain.

In the military sector, the following criteria always arise, irrespective of the product: Robustness, simplicity, durability, and ease of repair. In addition to general system robustness, the drone must be able to be used under restrictions, even in adverse weather conditions. Currently, the technical limits for sUAS are rain, hail, high wind speeds, and extreme cold. In other words, almost all conditions that become really strenuous for the infantryman. The drone must be simple in its design and handling. Reduced complexity also reduces the logistics behind it as well as the necessary technical knowledge, and virtually always increases reliability. At the same time, simple, intuitive operation reduces the amount of training time required. For cost-efficient and reliable use, the equipment must achieve the longest possible service life. As many spare parts as possible should be able to be replaced by the operator himself without tools, and compliance with technical deadlines and inspections should be kept to a minimum.

SUAS/NUAS DEPLOYMENT PROCEDURE

With all the enthusiasm for today's technical possibilities, leaders and operators must never forget: The deployment of a drone is just another tool to achieve the attack objective, not an end in itself. Every drone flight has a clear mission. Drone operations are therefore ordered after weighing all tactical factors against each other, starting at the squad leader level (in most situations, only with the approval of higher-level command). At a minimum, the drone squad must be given the following parameters: **Launch time** and **location**, **cleared flight paths**, **reconnaissance target** and **behavior at the reconnaissance target**, **landing time** and **location.**

A short combat order to the drone squad can be structured according to the acronym **"DR²ONE"** or **"DRRONE."** Let's put ourselves in the position of infantry squad A1, which is about to meet an enemy in the area *GRAMSCHATZER WALD* (a big forest around the German village Gramschatz). The squad leader gives the order to deploy the drone.

DEPLOYMENT (When? Where? For how long?)	"Drone team D, ready for takeoff, NLT 0445 in clearing near squad position, flying time max. 15
ROUTE (Where to?)	with departure near ground level, direction SW for 500, then N for 800, in succession at will.

RECONNAISSANCE (What? How?)	Recces dismounted infantry north of GRAMSCHATZ, reports type, strength, and behavior immediately, and breaks off reconnaissance immediately in case of defensive measures by the enemy.
EVASION	After landing, as soon as possible, move to the ordered change position.

What was the order in this example?

- **"Drone team D, ready for takeoff NLT 0445 in clearing near squad position, flying time max. 15"**

In the case of a drone team, readiness is established when the operator has prepared the launch site so that the drone can take off immediately on command. In open terrain, this includes a neutral surface (poncho, jacket, etc.) to keep sand and dust away from the device as far as possible. In rainy conditions, the team itself is protected from the rain in such a way that the screen can be clearly seen. The takeoff path must be free of branches, wires, and other objects, at least in the vertical direction. When the drone team is ready, it reports to the squad leader. In this case, the leader wants takeoff to take place at any time from 0445 on. He has already selected a suitable clearing in otherwise wooded terrain as the launch position. The current situation (battle command schedule, higher command constraints, or similar factors) allows a 15-minute window between launch and landing. While the drone is in the air, at least the drone team can hardly move because it is fully engaged. Meanwhile, it must also be secured, as it can barely keep an eye on its immediate surroundings. So the squad can not move on until the drone team can keep up. Otherwise, they risk loss of the mission equipment.

▶ **"with departure near ground level to SW for 500, then N for 800, in succession at will"**

The squad leader orders the drone team to, whenever possible, follow a specific departure path that is intended to disguise the current position of friendly forces. Although the reconnaissance target is to the north of his own position, he chooses to turn southwest behind a ridge. The final approach to the enemy will not be in a straight line. In this way, he can also coordinate the drone launch with neighboring groups, which might otherwise assume the sighting of an enemy drone. In this case, the path is only specified until this goal is reached, after which the drone team leader himself decides on the flight movements *("in succession at will")*.

▶ "recces on dismounted infantry north of GRAMSCHATZ, reports type, strength, and behavior immediately, and breaks off reconnaissance immediately in case of defensive measures by the enemy."

The reconnaissance target is the terrain north of GRAMSCHATZ where enemy movements are suspected. The drone team needs to confirm:

1. **Is the enemy currently there?** ----► ("north of GRAMSCHATZ")
2. **Is it really dismounted infantry, or are there vehicles too?** ----► ("type")
3. **How strong is the enemy?** ----► ("strength")
4. **What is the enemy doing?** ----► ("behavior")

Once the recce flight has provided answers to these questions, the team leader prepares a report on the enemy for the squad leader. This could read, for example:

"A1, this is A1D, enemy report: Between edge of woods and northern edge of village GRAMSCHATZ currently hostile dismounted infantry in company strength, widespread, suspected intent to penetrate village, I maintain line of sight."

The team leader supplements his enemy report with the information that he still has line of sight. This allows the squad leader to give a follow-up order if needed, e.g.:

"A1D, this is A1, copy. Order: maintain line of sight as long as possible, additional flying time cleared."

The follow-on order is to maintain contact with the enemy as long as possible. How long is that? Either until the reconnaissance has to be aborted due to the given maximum flying time *("15 minutes")* or until defensive measures are taken by the enemy (*"abort reconnaissance immediately if defensive measures are taken")*. Here, the squad leader has canceled the maximum flying time of 15 minutes. The drone team leader now has the whole flying time at his disposal.

DETERMINING THE MISSION VALUE OF INFANTRY DRONE TEAMS WITH SUAS

The mission value of a squad depends on the terrain, the carried weapons and ammunition, the supply situation, personnel strength, and the available friendly and hostile capabilities. In the following, we will consider the mission value of a rested and fully supplied drone team in different types of terrain. The adversary capabilities can be transferred from the following section **"Defense against sUAS."**

▶ URBAN TERRAIN [MAJOR CITIES AND METROPOLITAN AREAS]:

This already highly complex combat environment offers a myriad of opportunities for takeoff and landing. The distances are often very short, which greatly increases the effective reconnaissance time. The drone team can usually operate under full cover. Essential to urban drone flight is a strong and stable data transmission, which regularly challenges the squad's voice radio in such an environment. In the case of particularly heavy construction, the squad must therefore expose itself on rooftops or squares and requires a stronger backup component. In such a case, drone operations can easily tie up the entire squad.

At the same time, reconnaissance of enemy forces in an urban environment is more difficult and less accurate because you can only see what is moving outside the buildings. Forces inside buildings, sewers, and behind any kind of visual cover remain hidden. This is where the addition of **nUAS (Nano Unmanned Aerial Systems)** is of great value. Devices such as **Teledyne FLIR's "Black Hornet" Personal Reconnaissance System (PRS)** can penetrate through the smallest of openings and provide reconnaissance in hard-to-reach areas **(see Chapter 4)**. Dismounted infantry drone squads should therefore also have nUAS, but widespread deployment is costly at the moment.

▶ WOODS AND FORESTED TERRAIN (TEMPERATE CLIMATE ZONES):

In the mixed forests of temperate climates, there are always enough gaps available for drone operations. While in the winter deciduous forests offer a high probability of reconnaissance, almost like open terrain, dense coniferous forests are not suitable. With full tree canopies in summer, the reconnaissance possibilities beside paths, aisles, and clearings are severely limited. Extensive forest areas are hardly suitable for drone operations.

▶ FORESTED TERRAIN (TROPICAL CLIMATES):

Dense jungle virtually precludes drone use. There are few, if any, takeoff and landing sites, and the area of operation in this type of terrain is often very large. The use of sUAS in the jungle only makes sense when the terrain changes. Otherwise, large drones with appropriate range and armament must be used here in a grid search.

▶ MOUNTAINOUS TERRAIN:

In low-mountain terrain, sUAS can have very high operational value. Mountain troops can move much more quickly if detailed advance reconnaissance is available. Since positional options are typically somewhat more limited and there are few points that offer an overview of larger parts of terrain, directing accurate fire is of particular importance. Security, which is sometimes very costly in mountainous terrain, can also be greatly simplified. The added value of armed and unarmed drones in this terrain has been demonstrated by their successful use in the war over the Karabakh Mountains in the Caucasus.

▶ HIGH MOUNTAINS:

The extreme weather conditions, low atmospheric pressure, and sometimes long distances militate against the use of sUAS. In addition, conventional light infantry cannot be successfully employed in this type of terrain.

▶ ARCTIC AND SUBARCTIC TERRAIN:

The greatest technical challenge in these regions is the extreme cold. Flight time is significantly reduced. There is also a risk of freezing condensed water. Apart from that, there is nothing to prevent a successful mission in subarctic terrain.

▶ ARID TERRAIN:

For use in very dry and hot environments, dust protection of the drone is the first priority. Plastic surfaces and the remote control must be as resistant to UV radiation as possible.

▶ CULTURAL LANDSCAPE:

The design of the Central European cultural landscape can now be found on all continents. Smaller towns and cities are surrounded by many subdivided agricultural areas and smaller forest areas. The entire terrain is covered with infrastructure (roads, paths, power and water lines). Streams and rivers crisscross the land, and the topography is undulating or hilly. In such a changing environment, small drones can be put to great use. Since direct lines of sight are often broken by the varied terrain, these operational assets greatly enhance the force's reconnaissance capabilities. In early 2022, the Ukrainian Armed Forces successfully deployed drones of various types in such terrain in the greater Kyiv area.

CONCLUSION:

The sUAS drone team achieves the highest operational value in cultivated landscapes as well as in mountainous terrain. If it is supplemented by a nUAS system (small drones), its operational value can also be rated as high in heavily urbanized terrain. Whenever visibility from above is severely limited (urban or forest) or the climate becomes extreme (desert, arctic, or tropical), the mission value decreases. Small drones are therefore ideal for integration into light infantry combat, as these limitations also apply to this force.

DEFENSE AGAINST SUAS

If you are looking for ways to use drones, you must assume that they will also be used by the enemy. So what defensive measures can be taken, or rather, how will the adversary protect itself from our drones?

▶ CAMOUFLAGE

Force camouflage is a significant but troublesome issue. Only disciplined forces can take care of this on a daily basis, as required, over an extended period of time. However, anything is possible, even in the 21st century. Anyone who believes that simple infantry forces have no chance against modern air reconnaissance is fundamentally wrong. In the Korean War, the United States Air Force missed the movements of several Chinese divisions (!) despite intensive and high-quality air reconnaissance. The failure to detect dismounted, disciplined infantry units is still likely today: The smaller the force, the easier it is to hide. But how does one camouflage against drone reconnaissance? First and foremost, routes that offer upward visibility protection are preferred. Forests, covered alleys, or larger hedges are suitable. If you need to prepare for defense or remain in a sector for a longer time, the provision of visual protection is essential: Paths between buildings are covered with sheets, fighting trenches are hidden with camouflage nets, and positions in the depths of rooms are chosen. Dispersal is also camouflage and protection: By disentangling forces, the probability increases that only a small portion will be reconnoitered, creating a false situational picture for the enemy.

▶ IMMEDIATE ACTION DRILLS (IAD):

Standard procedures are established in training for what to do when a drone is sighted. Such predefined actions according to a certain scheme are also called Immediate Action Drills (IAD). By following a fixed procedure, one gives up any flexibility and possibility to adapt to the situation but gains a massive increase in reaction speed since there is no decision-making or commanding. The procedure is simple: The infantry squad is on the march to its staging area. An infantryman detects an approaching drone about 300 meters away at a low altitude. He calls out, "DRONE!," triggering a standardized behavior. The entire infantry squad breaks formation, and everyone jumps to the nearest position that provides upward visual cover. There they wait while the squad leader prepares a decision and gives an appropriate order for further action. This behavior is nothing new and is essentially the same as the behavior during an air attack. If necessary, the infantry squad has suitable defensive means that it can bring to bear. A standard procedure can be established and recalled for this case as well.

▶ DEFENSE MEANS:

Portable drone defense capabilities are currently severely limited. Shotguns with special projectiles represent a rudimentary variant for the immediate area: Instead of buckshot, the case inherits metal pieces with strings that unfold into a net after firing, causing the entangled drone to crash. Classic bird-shot can also be used to achieve kills. This solution is only useful for cluttered, preferably urban terrain; Distances are very short there, and infantry should carry shotguns in this environment anyway. Some defenses in civilian use are not suitable for the infantry.

These include **trained birds of prey, fixed jamming devices** ("jammers"), or **dedicated interceptor drones.** Portable jamming rifles are currently the only option usable for these forces and are specialized for the function. For the time being, the shotgun probably remains the only realistic defensive weapon for dismounted light infantry; It is questionable whether the potential utility of a specialized jamming rifle really justifies its additional weight.

The **shotgun,** which will be carried anyway, and good camouflage training weigh considerably less. It is conceivable, however, that in the future infantry operations will be more heavily supported from the depth by suitable air defense forces and that they will take over the fight against small drones as well. Mechanized forces can already take on such a mission with **long-range machine cannons** and associated **air-burst munitions**, or even with new types of **lasers**. Dismounted forces will thus tend to rely on the availability of vehicle-based solutions for active defense in the foreseeable future.

3 TYPES OF THE MOST OFTEN USED COMMERCIAL DRONES

"Drone" can mean many things. From the palm of your hand to the size of an airplane, anything is possible. Some have wings, others are quadcopters with completely different flight characteristics. The latter have gained acceptance on the civilian market. Rotary-wing drones are the devices that are proving particularly useful as tactical reconnaissance vehicles.

This is because helicopter-like rotary wings enable these aircraft to take off and land vertically or hover to keep a target in sight at all times. This design is especially popular for consumer drones, but their capabilities are also useful on the battlefield. These drones tend to move more slowly than fixed-wing aircraft and have shorter flying times, so troops are more likely to use them for short-range missions.

DJI MAVIC

Operators: Ukraine and Russia | Manufactured in: China | Wingspan: 12 cm | Flying time: up to 30 min | Top speed: 60 km/h | Engine: electric | dji.com

The "Mavic" series quadcopters made by the Chinese company DJI are the most commonly used drones in the conflict and are operated by both sides. At the beginning of the war, the Ukrainian Defense Ministry asked civilian drone owners to donate their "Mavics" for the war effort; thousands more came from foreign supporters.

One of the smallest versions is the "Mavic Mini," which weighs only half a pound and folds up small enough to fit in a carrying bag. Despite the drone's tiny size, the electronics of the "Mavic Mini" are powerful enough to send high-quality videos over a distance of 3–4 km. This makes these drones handy tactical reconnaissance vehicles in urban areas, where Ukrainian forces use them to locate Russian vehicles and prepare ambushes for tank-hunting teams with "Javelins" and other weapons. Individual snipers use them to locate targets, and artillery units deploy "Mavics" when nothing larger is available.

AUTEL EVO II

Operator: Ukraine | Manufactured in: China | Wingspan: 40 cm | Flying time: 40 min | Top speed: 100 km/h | Engine: electric | autelpilot.eu

Most drones flown in this conflict are not manufactured by military contractors, but are commercial off-the-shelf quadcopters. The "EVO II" by the Chinese company Autel has a maximum takeoff weight of 4.5 pounds, making it one of the largest consumer drones available. Both sides use the "EVO" for local reconnaissance and tactical bombing. Ukrainian forces and Russian-backed separatists in the Donbass region use the "Evo" to drop so-called Chattabka munitions—small 30 mm VOG-17 shells that eject deadly shrapnel within a radius of up to six meters around the impact site. There is another reason Ukrainian operators like the "EVO:" It is reportedly highly resistant to Russian jamming attempts.

TEAL GOLDEN EAGLE

Operator: Ukraine | Manufactured in: USA | Wingspan: approx. 50 cm | Flying time: up to 55 min | Top speed: 80 km/h | Engine: electric | tealdrones.com/suas-golden-eagle

After the U.S. Army banned its forces from using Chinese-made quadcopters due to security risks (the drones' radio controls are unencrypted, and the devices could potentially capture and store sensitive information that would be shared with the Chinese government), the U.S. began developing its own alternatives as part of a defense program called Blue sUAS. Some of these alternatives, including the "Golden Eagle," were sent to Ukraine.

The "Golden Eagle" resembles a quadcopter for private use but meets military standards and features secure, encrypted communications and advanced computer technology. It has a high level of autonomy, made possible by sophisticated features such as visual odometry. This system scans the ground to calculate speed, distance, and direction, allowing the drone to navigate accurately even when GPS is jammed. It is also equipped with a high-quality thermal imaging system that can be upgraded as technology evolves, according to the Army.

SKYDIO X2

Operator: Ukraine | Manufactured in: USA | Wingspan: 66 cm | Flying time: 35 min | Top speed: 40 km/h | Engine: electric | skydio.com/skydio-x2

The “Skydio X2” is another American drone built according to Blue sUAS standards in response to data security concerns about Chinese-made models. An AI-controlled engine allows the consumer versions of the Skydio to fly autonomously; you can set it to follow mode and take videos while you drive down a mountain or walk along a beach. The military version has a few more features, such as integrated obstacle avoidance, encrypted communications, and a thermal imaging camera for nighttime use. Skydio has donated dozens of drones to the Ukrainian Defense Ministry and sold hundreds more to groups that support Ukraine, reportedly passing them on to the country’s Aerorozvidka drone flight forces.

But this can be categorized beyond ordinary users, with high costs and sophisticated uses. The tactical drone, which is useful for everyone, is small and handy and can be used as a scout or reconnaissance aircraft. The first mass-produced type is the “DJI Mavic Mini” mentioned above. Especially useful are versions from Mark 2 on, because no modem is needed for control. There are counterpart models by other manufacturers, such as Autel, e.g., the “EVO Nano.”

Both are available in a “Fly More Combo” (DJI) or a “Premium Bundle” (Autel). These are handy, professional devices for civilian users that can also fly reconnaissance missions well. The Autel drones seem to be somewhat less susceptible to jammers and are definitely more independent of the parent company **(see Chapter 8).**

Beyond that, at the lower end of evolution in the modern drone world, there are plenty of commercial models that also fly stable and transmit images of varying quality. If you simply type the search term “drone” into an online trade platform, you will see the full range. Generally speaking, you get what you pay for in this product category as well. But this does not mean that smaller, less expensive drones cannot be good close-range reconnaissance devices. You just have to know their limitations: The main considerations are usually a less stable flight behavior and compromises according to the used optics. In principle, these drones are tactical consumables. And always remember: It’s better to have a cheap drone than no drone.

PD-100 BLACK HORNET PRS	
Rotor diameter	12 cm
Fuselage length	10 cm
Fuselage width	2,5 cm
Range	1500 m
Weight	18 g
Maximum speed	36 km/h

WORKING WITH PROFESSIONALS: BLACK HORNET PD-100 PRS

Norwegian manufacturer Prox Dynamics has been offering a nano reconnaissance drone for civilian and military use since 2007. The company achieved its commercial breakthrough in 2012 with the first sale of larger numbers to the British Army. In the same year, a total of 65 of these systems were deployed in Southern Afghanistan. Since then, the production of the **"Black Hornet"** has been ramped up. Users include special operations and infantry forces worldwide. The latest procurement decisions will make Norway the largest user in the short term. Right from the start of development, the Norwegian Armed Forces repeatedly provided unbureaucratic support for conducting trials and field tests. In 2012, a smaller number of units were procured for testing purposes by the Norwegian Special Forces and the Telemark Battalion. Thus, while serving in this unit, the author was given the opportunity to use one of the first prototypes in training and exercise.

The PD-100 PRS (Personal Reconnaissance System) is delivered as a package with a controller and two rechargeable drones. The total weight is 1.3 kg.

Since the charging time is equal to the flight time, constant use is possible as long as the "Black Hornet" can be charged. Only the approach and departure times to the target represent reconnaissance gaps. The drone flies with minimal noise and visual signature at up to 30 km/h up to a distance of 1.5 km. It can be used in rain and at wind speeds up to a maximum of 12 m/s (maximum).

The author experienced the use of this system as a potential game changer in decision-making.

During a two-party exercise with the United States Marine Corps at a Norwegian urban combat facility, the author was deployed as a squad leader in a reinforced Norwegian infantry company. As part of a strongpoint defense planning, the objective was to deny the enemy battalion the capture of several buildings that dominated the surrounding terrain. The author's mission was to conduct one of three different pre-planned counterattacks, which were to be initiated depending on enemy behavior. In each case, the counterattack took us down a side street into one of three different multistory buildings. The terrain behind those houses could not be seen. Only a section of the area could be monitored by a detached sniper team. When the snipers reported enemy movement, the platoon leader in the sector needed more information to lead the proper counterattack; an enemy deception was possible. The author ordered the deployment of a “Black Hornet” and followed what was happening in real time via the drone soldier’s control unit. Although the PRS was not yet night-combat capable at this point, insight into the partially dark rooms was sufficient to detect movement. Marines had already penetrated one floor of the building on the other side of the street. By flying around the twisted structure, the point of entry could be identified. The opportunity was favorable: The Marines took the building via two ladders, and a typical “traffic jam” occurred at this point. The drone mission was immediately aborted, and the counterattack was launched directly at the breach site. Two Marine squads were now cut off, and the next attacking subunit could be crushed in a surprise ambush with rifle fire. Thus, the use of a “Black Hornet” significantly influenced the defensive success of a Norwegian company. The “enemy forces” did not learn how they had been reconnoitered until the debriefing; no one had noticed the drone.

LESSONS LEARNED:

- The sensorial overload in combat ensures that the "Black Hornet" is only reconnoitered by chance, if at all. It is simply too small and too quiet.

- If it is detected, it can only be engaged at arm's length and with a quick reaction. Targeted jammers have been unsuccessful so far; only large vehicle solutions offer protection here, but this also limits one's own means.

- If a "Black Hornet" is lost, nothing is lost. The data is transmitted, not stored on the nUAS. The drone cannot be traced.

- Due to its small size, it can be easily deployed anywhere and is not subject to any restrictions regarding a launch site.

- Due to the possibility of automatic flight based on waypoints, it is ideally suited to replace a patrol in urban terrain. It saves forces and is more quiet. The operator simply stays in his sleeping bag.

- Urban terrain is obviously the most favorable location for use, but the drone has also proved very useful in forest combat. Advance reconnaissance over some 500 to 600 meters is incredibly valuable for dismounted infantry at visibility levels below 100 meters.

DEVELOPMENT OVER THE PAST TEN YEARS

Since 2014, the drone has also been capable of night vision. The manufacturer Prox Dynamics was bought by the U.S. company Teledyne FLIR in 2016 for $134 million. The variant offered since 2022 is once again of higher quality in terms of flying time and materials used, with a significantly lower price at the same time. Between 2015 and 2022, the acquisition costs fell from almost $200,000 to around $50,000 due to the expanded serial production.

More information on the "Black Hornet":
flir.com/products/black-hornet-4/

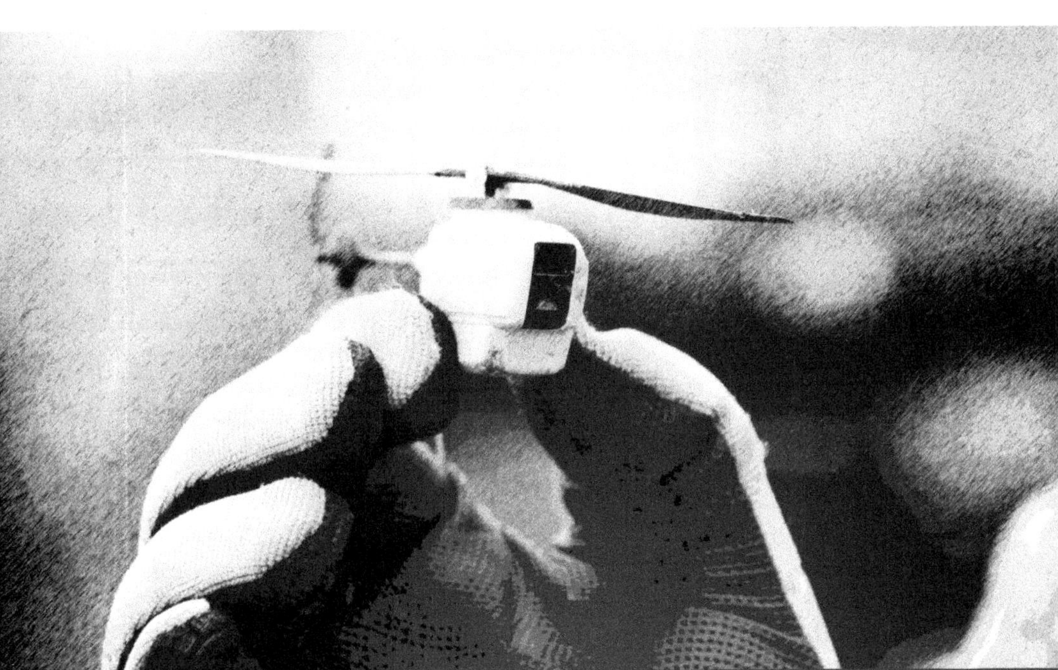

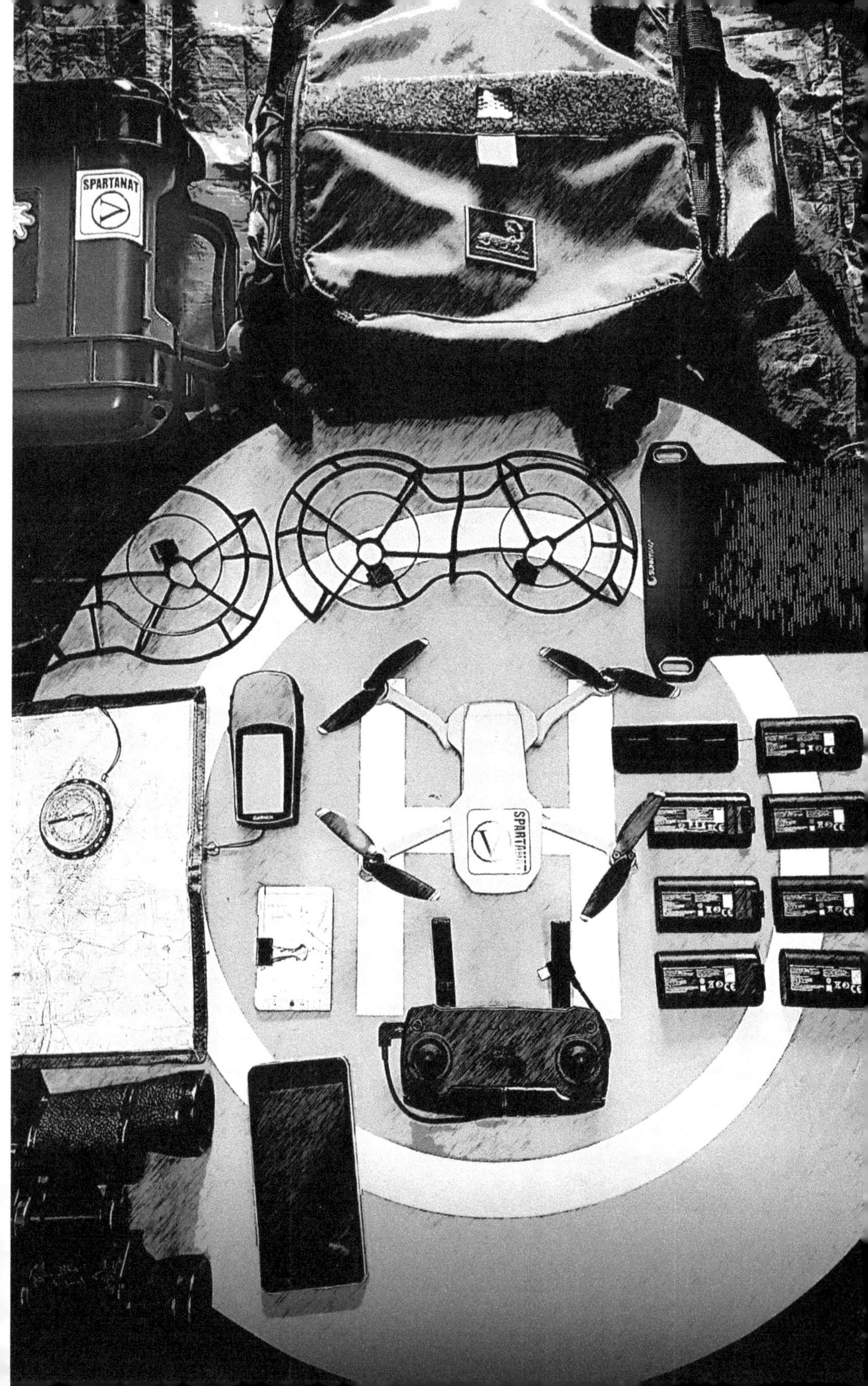
SPARTANAT

ADDITIONAL MISSION EQUIPMENT

Classic civilian drones that are useful in a tactical environment usually come in small (drone, controller, and battery) or large sets (drone, controller, multiple batteries, and bag). In the case of DJI, the latter is called a "Fly More Combo." This only refers to the additional three batteries included. There is quite a bit more that can be useful for the tactical drone.

▶ CASE

A broken drone is of no use. DJI includes cases or padded bags with its "Fly More Combos." Some of these are fine and certainly sufficient for civilian use. For tactical applications, only the waterproof hard shell case counts, which is available primarily for the popular DJI models. The hard shell guarantees that nothing will happen to the drone during transport, no matter if the case falls down, someone drops another piece of equipment on it, or other forces act on the case. This is achieved through precisely cut foam inserts that protect the system from shocks.

▶ SPECIAL BACKPACKS

A backpack the size of a one- or three-day pack is enough to carry a drone and everything else you need in the field. Models for special equipment like sniper gear, cameras, or medical equipment can be converted into backpacks for drone operators. The most practical packs will be front loaders. External solar panels can be beneficial for charging the drone's batteries.

▶ DRONE LANDING PAD

In general, it is not necessary to buy special drone landing pads. Jackets or any other heavy fabric and a few stones will do the job for small commercial drones. Otherwise, landing pads exist in all sizes and shapes. They are of great use in extreme environments, especially in the desert.

▶ SOLAR PANEL

Independence in the field also means independence from the power outlet. In other words, it cannot be guaranteed that there will always be one nearby and that it will work when electricity is needed. The solar panel is a nice thing to have, even if batteries can normally be charged in the vehicle. The range of models offered is wide, and the price is not always decisive. Ideal is a panel that can be attached to the backpack. With it, the batteries of the drone or smartphone can be charged even during rest periods.

▶ POWERBANK

A powerbank is an energy storage device that you can carry in your pocket. It is worthwhile to always have one charged in your backpack as a last resort. It can also be two or three. There are big differences in prices here. Inexpensive models also do their job, a plastic bag makes them waterproof. Outdoor models exist too, which are more robust and waterproof to some degree. Checking them regularly for charge is important, especially when the cold season takes its toll.

▶ SUNSHADE

Normally, a smartphone is plugged into the controller as a screen. When the sun is at certain angles, the observer does not see anything. There are sun visors that cover the phone so that everything can always be seen clearly. These can also be improvised.

▶ FILTER

For better documentation quality, there are ND and polarizing filters. ND filters (neutral density filters) are basically sunglasses for your camera. In strong sunshine, this allows you to increase the exposure time without overexposing the film stock. Polarizing filters can take out reflections, but only if they are aligned correctly. Unfortunately, readjustment on the fly is not possible. If you want to create truly cinematic footage, you can't get around filters. For mere reconnaissance, you can get along just fine without them.

▶ NIGHT FLYING LIGHTS

For some drones, there are kits with LEDs that work like position lights. These are useful for night training flights and ideal for civilian use as well. Visibility, of course, is not a desired feature for tactical drones. For indoor use, spotlights on the drone increase observation capabilities.

▶ PROPELLER PROTECTION

For many drones, there are propeller protection devices available that can be clipped to the rotor arms. It prevents the rotors from smashing into surrounding objects and avoids damage. On the one hand, this is practical, so that beginners do not have to replace the propellers on a continuous basis. On the other hand, propeller protection is essential when the drone is used for tactical indoor reconnaissance or in an environment with many nearby obstacles. It is therefore worthwhile, especially for close-to-the-ground operations or in built-up terrain.

▶ SPARE PROPELLERS

No propeller means no flying. So you not only have to take care of your propellers, but also have to carry enough spare parts. Besides the original parts, there are also many derivative products. Always try them out before using them on a mission. On top of that, there are propellers that claim to be particularly quiet, which helps in the desired quiet approach. Whether such propellers—expensive and sometimes made of titanium—are of any use for your model, you have to find out yourself.

▶ TOOLS

Reserve propellers in the backpack do not mount themselves, so you have to take with you the right wrench, which is usually supplied with the drone. A backup wrench or a small tool kit with matching bits and a screw set should be standard.

▶ CAMOUFLAGE SKINS

Drones come in a variety of colors. The common models from DJI are mostly light or dark gray. This is not bad at all; especially when you look from the ground against the sky, it is very inconspicuous. Whoever thinks that the drone needs camouflage better suited for changing environments can improvise with all sorts of means. Colored pencils work just as well as ready-made skins that fit the drone model precisely. They come in olive, woodland, multi-cam black, and many other patterns. On **etsy.com,** there are endless variations.

▶ SD CARDS

Memory is essential if you want to keep the footage. Here, it pays off to be on the safe side and use the more expensive, established models from SanDisk. Cheap, no-name products from China can also work, but they don't have to. Again: "One is none and two is one." The memory size of the cards depends on your drone. The manufacturers usually state the maximum size of memory cards that are supported. It also depends on the resolution you are recording at. Video recordings in 4K use considerably more storage space per minute than recordings in 1080p.

▶ BATTERIES

Batteries are the drone's gas station. Regular drones usually come with one piece, DJI's "Fly More Combos" come with three. 20 to 30 minutes of flying time per battery is the standard. But one is none. Carrying five to ten batteries is certainly not a mistake. Check the charge levels regularly, especially in the winter, so that there are no unpleasant surprises!

▶ MAPS AND GPS

It may be enough for you to use the tactical drone to observe what is currently visible and pass it on directly. However, it may also be useful to check out the area where you are working. Electronic aids such as a GPS device (Garmin, etc.) with or without maps are handy. When in doubt, a classic map and compass will work regardless of any power failure or device damage.

▶ RAIN COVER OR PONCHO

Rain is the enemy of the drone. It does not like it in the field. Even if nothing happens at first, water will harm it in the long run. So it is worth protecting the drone backpack. Either with a rain cover, which prevents the valuable equipment from getting soaked, or with a poncho, which protects the man and the equipment and can be used as a protective cover or even as a drone landing pad after taking it off.

▶ ALTERNATIVE CONTROL APP

Drones can sometimes do more than their manufacturer lets you know. As an example, consider the DJI "Mini" or "Mini 2." These can not perform a self-guided "follow" or fly off predefined waypoints out of the box. With alternative apps, such as "Litchi," this is possible. This way, a predefined path can be flown via GPS, including camera recording, without you having to steer it yourself. When "following," make sure that your drone does not fly into obstacles because it has no obstacle detection. If you decide to use an app like this, be warned: The warranty expires immediately. Of course, you will still have to comply with legal requirements.

ORIENT YOURSELF!

Although drone operations are three-dimensional, their targets are almost always on the ground. To perform a successful mission, you need detailed knowledge about the terrain. Unless you are familiar with it because you call it home, the classic way to do this is with a **map** and **compass**. The map gives an overview of the terrain and is independent of any electricity source. Additionally, as a non-electronic device, it cannot be located. Together with the compass and a good knowledge of what to do with it, it is a valuable tool for the drone operator. This should be practiced by every drone team.

In the age of **GPS**, however, electronic means of orientation are of great use. Garmin helps here efficiently with its products. Those who want more military utility can also be helped with software that runs on smartphones: “ATAK” comes from the U.S. and turns the phone into a tactical command center.

ATAK

"ATAK" is an app for Android smartphones. The abbreviation stands for **Android Team Awareness Kit**. It enables you to navigate accurately and exchange data with others. You can also share Points of Interest (POI) with others in real time and see your own location and that of your colleagues on the map. This sounds great, but in order to fully use "ATAK", the user needs a lot of plug-ins, downloaded maps and special radios (with additional plug-ins) to share the data with others. The civilian version of "ATAK" is called **"ATAK-CIV"** or simply "TAK", but in the end it is the same program. It lacks all the plug-ins, but some of them can be installed later. Recently, there has also been an "ATAK" version for iOS devices available, which is called **"iTAK"**. However, this version is somewhat different in appearance than "ATAK". "ATAK-CIV" is available for free from the **Google Play Store**; "iTAK" can be downloaded for free from the **Apple App Store**.

Official website: **tak.gov**
Community site around the civilian version: **civtak.org**

TACTICAL NAV

An alternative to "ATAK" is the app **"Tactical NAV"**. This paid app comes with everything you require for orientation in the field. It is easier to use than "ATAK", but also offers fewer functions. As an accurate map and very precise navigation tool, "Tactical NAV" is even better than "ATAK" in our opinion. The only drawback is that the map material is downloaded via the mobile data connection during use. "Tactical NAV" is available for iOS and Android and can be found quickly in the Play Store and App Store. The version for Apple is slightly better than the Android version in terms of operation and range of functions.

Official website: **tacticalnav.com**

THE TEN COMMANDMENTS FOR THE TACTICAL DRONE OPERATOR

Anyone who has the opportunity to put drones into the field should be as prepared as possible and keep a few things in mind:

1. TRAINING AND PRACTICE ARE HALF THE BATTLE.

Professional initial training will save you a lot of time and money. After that, practice time counts—the more, the better. None of us are born drone pilots. As with any piece of equipment you use, it's the practice that makes the survivability. Do you have the time and fun to fly the drone? The more you do it before it gets serious, the longer you will last in the field. The only way to exploit the potential of your system to the maximum is to master all facets of drone flight.

2. YOU HAVE TO ACQUIRE DECENT SKILLS IN OBSERVATION.

Evaluating aerial imagery works differently than observation with a direct line of sight from the ground. Learn to spot camouflaged objects and people. Using the camera on the drone also requires practice; it is your eye. Learn to look for the right things. Practice from a variety of angles and positions.

3. KEEP YOUR SIGNATURE LOW.

The drone's flight sounds, movements, and reflections can give it—and perhaps you—away. Learn under what conditions you can be heard and from what distance. The noise can give away the flying drone. Pay attention—with and without wind, reflective house walls, etc.—so you learn to stay quiet and not give away your position to the target of observation.

4. IMAGINE THE VIEW OF THE ENEMY.

From which distance can you spot it? The biggest self-protection for your flying eye is not being seen. Also, try observing the drone with night vision or, even more exciting, thermal imaging. Then you will know for sure about its heat signature. And never give away your position by coming straight back! Never fly back on a direct route, make a loop instead that cannot be seen. You can also try a deception flight in the form of a fishhook. When they think you are gone, they will come out of their cover.

5. ADAPT YOUR FLIGHT PROFILE TO THE GIVEN TERRAIN.

Try to take hidden or unexpected routes. Ancient wisdom, eternal principle: No one is invisible. If the drone is observing, it is observable.

6. BE AWARE OF YOUR IMMEDIATE SURROUNDINGS

You are vulnerable when operating the drone. You are busy with the drone? Furthermore, you're flying it all focused and have spotted something? Pay attention to your immediate surroundings. But most importantly, make sure you have one or more buddies nearby, as long as you keep an eye on the screen.

7. THERE ARE NEVER ENOUGH BATTERIES.

When you need a full battery, it will be dead. If you want to fly, you require fuel. And enough of it, because, as we all know, the battery runs out when it is most needed. Like with magazines, there is no such thing as too many batteries. Beware of cold conditions, keep your batteries close to the body in alpine or arctic environments.

8. KNOW YOUR ENEMY. THEREFORE, FLY FAST AND ACCURATELY.

The best protection against jamming and drone defense is speed. So learn to generate the image you need in the shortest time possible. The reconnaissance drone will always disturb the enemy. The enemy wants it destroyed. He will shoot at it, jam it if he is prepared, or jam the GPS signal. Deal with this important issue. Ideally, you gather first-hand experiences about the jamming reaction of your drone.

9. MACHINES NEED MAINTENANCE.

Maintain, service, and repair your drone yourself. Carry spare parts and avoid damage. The drone is a valuable piece of equipment because it gives you special abilities. But it is technology. Take care of the device. Have a second one with you, if possible. Know how to repair the drone if necessary. If it is beyond repair, a broken drone can provide spare parts for the next one.

10. IN PEACE, THERE ARE RULES, IN WAR, THERE ARE NONE.

We fly the drone in peace. Obey the laws. Respect the flight zones and the prohibitions. In an emergency, these rules no longer apply. Then use the drone in the way that is most efficient.

CIVIL DRONE REGULATIONS

7

U.S. DRONE LAWS [EFFECTIVE 2024]

Rules and regulations regarding drone operations in the U.S. and UK are constantly evolving. It's best to check with the Federal Aviation Administration (FAA) in the U.S. and the Civil Aviation Authority (CAA) in the UK for the most up-to-date information.

- Compliance with FAA regulations is mandatory for all drone operations in the U.S.
- All drones must be registered with the FAA and marked with the registration number.
- Pilots must pass an aeronautical knowledge test and obtain a Remote Pilot Certificate.
- Pilots wishing to work commercially must obtain a Part 107 authorization.

- Drones must be operated within the pilot's visual line of sight and not above 400 feet above ground level.
- Drones must not be flown over restricted airspace, near airports or manned aircraft, or in a careless or reckless manner.
- The FAA has established a number of safety regulations, including, but not limited to, the maintenance, inspection, and repair of drones.

► **All current regulations: faa.gov/uas**

UK DRONE LAWS
[EFFECTIVE 2024]

1. Drone operators must be at least 12 years old to fly by themselves.
2. Drones are not permitted to fly above 400 feet.
3. Operators must maintain a line of sight with their drones at all times.
4. Permission is required before flying in restricted airspace.
5. Do not fly your drone within a 3-mile radius around airports.
6. A minimum distance of 165 feet must be maintained from uninvolved persons.
7. Drones lighter than 0.55 pounds are permitted to fly closer to and over people.
8. Drones heavier than 0.55 pounds must be operated at least 500 feet away from parks, industrial areas, residential zones, and other built-up locations.
9. If a drone is equipped with a camera, the operator must apply for an Operator ID with the CAA.
10. Insurance is mandatory for the commercial use of drones.
11. Compliance with these regulations is required during both daytime and nighttime operations.

▶ **All current regulations: dronesurveyservices.com**

EU DRONE LAWS (EFFECTIVE 2024)

As of January 2021, the European Union Aviation Safety Agency (EASA) has implemented new regulations on drone management across the European Union (applying to all 27 EU member states, Iceland, Switzerland, Liechtenstein, and Norway). The goal is to provide a unified legal structure for drones operating in those countries. Previously, the legal framework for drone operations across the EU was highly fragmented, with each country adhering to its own rules and regulations. Experts believe that a unified model will streamline the regulation process, thus making drone operations easier and safer for everyone.

NEW DRONE CLASSIFICATION REGULATIONS

The new regulations stipulate that all drones must be classified into three categories: Open, Specific, or Certified. Along with risk factors, these categories factor in drone weight, certification, qualification of the operator, and operational features. For a smooth transition, EASA has decided to use the "Limited Open Category" during 2021 and 2022. This gives users ample time to get used to the new policies.

During the transition period, leisure drone operations should normally be classified into this category. According to the category guidelines that will be in place in the current two-year period, drones lighter than 4.4 pounds can fly as close as 165 feet to bystanders. However, drones heavier than that must maintain a minimum horizontal distance from bystanders of 500 feet. Their operation also requires the pilot to undergo an online training program as defined by the respective National Aviation Authority (NAA). See the table below (Fig. 1) for more detailed information on the Limited Open Categories.

Additional information on the Open Category for civil drones can also be found on the EASA website. The image below Fig. 1 displays a sample Open Category license as issued by the Austrian NAA as proof of completion of the online training program.

▶ **All current regulations:**
easa.europa.eu/en/the-agency/faqs/drones-uas

WHAT DRONES ARE ALLOWED TO DO IN THE EU

EU Drone Regulations divide the operation of drones into three categories. A distinction is made between "open," specific," and "certified" according to weight and operating environment.

CATEGORY "CERTIFIED"
(authorization required)
flight altitude above 120 m

CATEGORY "SPECIFIC"
(authorization required)
flight altitude above 120 m

CATEGORY "OPEN"
(no authorisafoptaluired)
maximum flight altitude 120 m

Drones <250g take-off weight

A1
close to people

A2
safe distance to people

A3
far away from people

A1 and A3: lateral distance of at least 150 m from residential, commercial, and recreational areas.

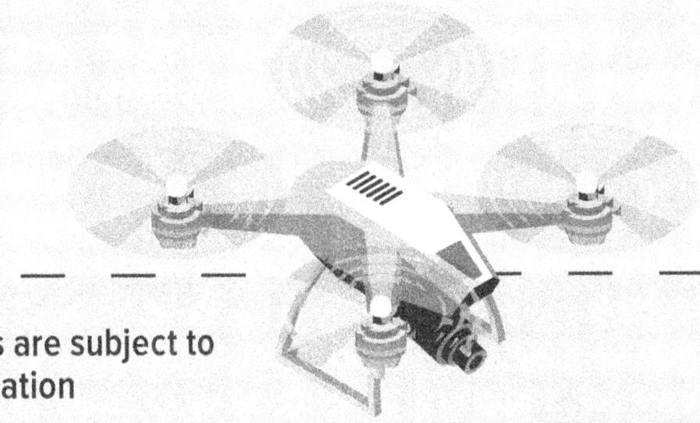

Drones are subject to certification

Drones are subject to certification

A GENERAL FLYING BAN APPLIES IN THE VICINITY OF AIR-PORTS OR MILITARY FACILITIES!

MAXIMUM PERMITTED FLIGHT ALTITUDE 120 METERS

In general, there must always be a direct line of sight to the drone.

MODEL AIRFIELD

unaffected by the regulation, except for labeling requirements

A1 and A3: Small EU Drone License

A2: Large EU Drone License (Remote Pilot Certificate)

A2: lateral distance of less than 150 m from residential, commercial, and recreational areas.

GEOFENCING: THE FENCE FOR DRONES AND HOW TO GET OVER IT

WHAT IS GEOFENCING? BASICALLY, IT IS A NO-FLY ZONE OR WARNING ZONE, IN WHICH YOU ARE ONLY ALLOWED TO FLY WITH YOUR DRONE IN A RESTRICTED MANNER OR NOT AT ALL.

Mostly, such zones include the vicinity of an airport or the area around prisons and military infrastructure. In these areas, it makes sense that drones may only be operated to a limited extent or not at all. However, not all manufacturers implement geofencing in the same way.

Companies like DJI have been incorporating geofencing into their devices since the first discussions about restrictions. However, other vendors, such as Autel, have taken the other side and do not implement geofencing in their aircraft. Only time will tell if this will continue in the future. Other companies, such as Parrot and Yuneec, have been using geofencing in limited form in their models for some time.

Geo-fences are created using tracking technologies such as GPS, RFID, Wi-Fi, and, of course, the drone's software. The "inner fence," so to speak, is part of the programming found in most drones. DJI products in particular use the drone's GPS receivers to automatically enforce airspace restrictions based on where the drone is located. This uses a digital airspace map that includes no-fly zones and other areas where drone flight may be restricted.

If you fly your drone into one of these areas, it will stop following you mid-flight, stop climbing, or simply refuse to take off. This can be an extremely frustrating moment for drone pilots. This feature was developed in the early 2010s by leading drone manufacturer DJI in collaboration with location partners. DJI unveiled the first geo-fence in 2013, which was turned into a final version in 2015. The concept has been adopted by other manufacturers as well as other tracking companies that provide GPS location data for drones.

In most cases, when you encounter a geo-fence, all you have to do is confirm that you are allowed to fly in that area. In other cases, additional, elaborate steps must be taken to request clearance.

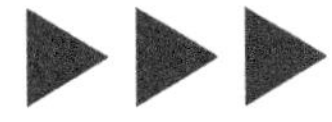

BUT IS GEOFENCING REALLY NECESSARY?

This is a fundamental discussion in which two camps have formed. Mostly, it's about the buzzword "security."

Permits are required for flights in sensitive areas (e.g., near airports). Various apps and some websites provide you with specific information about drone flights in these areas, such as flight altitude restrictions and the like. Geofencing, on the other hand, also refers to these types of areas but does not have the same information. The area is either restricted or unrestricted.

So the boundaries that are set in the airspace permit are not the same as a geo-fence. Every flight a drone pilot undertakes should be thought out and planned. Even flights around your own house should be well-thought-out and planned. **Once again, it's all about safety.**

The question now arises: Should everyone always be responsible for their own actions, or would it be better to have tireless technical aids such as geo-fences, which somewhat incapacitate the pilot but in return, for example, prevent an airport blockade and guarantee safety?

TYPES OF GEOFENCING

We look at DJI as an example here because it is the dominant company in the market and has an established geofencing system.

DJI uses a tiered zone system. Of the zone types, only two require the additional steps of unlocking. These two zones are the authorization and flight restriction zones. All other zones can be accessed onsite with a confirmation click in the app.

AUTHORIZATION ZONES

An authorization zone is represented by the **color blue** on **DJI geofencing maps.** Flying within such a zone is **not allowed by default.** If you are in one of these zones, the drone cannot take off without first unlocking the zone. **Unlocking** is done by entering the flight controller's serial number on the DJI website. To obtain the unlock license, the smartphone, drone, and controller must be connected to an active Wi-Fi connection, and you must be logged into your DJI account.

FLIGHT RESTRICTION ZONES

Flight restriction zones are shown in red and generally exist at landing and takeoff runway ends and over high-security facilities such as government buildings, prisons, or military installations. For such restricted areas, you must provide DJI with documentation from the appropriate authority to clear the zone. Other than that, clearing the zone is not much different from clearing an authorization zone (see above). In this case, you will need an airspace permit from the aviation authority, which you will need to enter on the unlock website to unlock the zone.

HOW DO I BYPASS GEOFENCING?

The only legal way is to buy a drone without geofencing, e.g., from Autel. At DJI and other manufacturers, you can unlock your drone as described above. Of course, this involves a certain amount of effort.

Using other control apps, such as "Litchi," does not override DJI's geofencing. That's because it's built right into the drone's firmware. The only way to really get around this is to hack your drone. There are dedicated forums and websites on the internet for this (keyword: "NFZ Mod"), but it's not that easy. Some models require soldering in addition to digital hacking. It might be easier to get a model that does not have geofencing from the factory and flies safely under your complete control—in peace as well as in action.

GEOFENCING IN A CRISIS [E.G., UKRAINE]

What does geofencing mean in a crisis or war, like in Ukraine? For the time being, nothing will change, since the geodata is permanently stored on the drone. They would only change after a firmware update. This means that even if the drone manufacturer imposes a no-fly zone over an entire country, the drone must first be updated for this zone to take effect. So as long as you don't update your drone's software, nothing will change on the aircraft.

For authorization and flight restriction zones, it is still possible to apply for an exemption as long as there is still access to the Internet in the regions in question.

More piquant is whether the GPS works, as the better drones all use it. This is because they are so easy to control and position. If the GPS fails or is disturbed, flying becomes more difficult because the positioning no longer works as precisely, and flying becomes more demanding. Again, if you've acquired practice beforehand, you'll be able to handle it.

THE DRONE AND ITS ENEMY: JAMMING

DRONES THAT ARE HELPFUL TO ONE SIDE ARE DISRUPTIVE AND DANGEROUS TO THE OTHER.

Responses to this challenge vary and are all still experimental. In the case of close-range nano-drones, it may be a hand strike that kills off the too-close reconnaissance aircraft. But that's probably a happy isolated case. At the other end of the spectrum, in use against larger military drones, are complex jammers designed to prevent successful deployment. We are interested in the range in between, and especially in small, commercial UAS. These can appear in annoying numbers, watching us from all sides—or-roaming the area as dangerous bombers.

Pragmatically speaking, the classic shotgun is a competent answer that is said to have taken down flying observers beyond the garden fence in the civilian world as well. Flying targets, admittedly, are not so easy to hit unless you have a serious degree of experience. In the civilian world, there have also been experiments with net guns that "catch" the drone and force it to the ground. Tactically, this is pointless because no one wants to carry around a net gun until it is needed. A very efficient solution involves lasers mounted on military vehicles that set the drone on fire. But until they have technically matured and are in widespread use, there are other efforts. The two most common are jamming the drone and disrupting the GPS signal. Both work best when used simultaneously.

Jamming is rather illegal for in civilian use. **In military and conflict situations, it is a different story: What works will be fine.** When jammed, however, the drone does not necessarily freeze in the sky, nor does it simply drop like a stone.

In simple terms, a **drone jammer** is a device that transmits electromagnetic noise on radio frequencies to superimpose those radio and GPS signals that the drone uses for operation. The frequency of a drone jammer is usually at 2.4 GHz or 5.8 GHz, which are public frequencies not intended for manned aircraft, public channels, or cell phone signals.

Jammers often look similar to futuristic rifles and work by projecting the jamming signal in the shape of a cone at an angle of 15° to 30°. First, the radio frequency signals to and from the drone are scanned in time and direction and then the jamming signal is sent. This is just "noise" that causes the drone to lose communication with its pilot. When this happens, most commercial drones automatically fly back to the pilot's location. Additionally, the drone's positioning system (GPS) can also be disrupted. In this case, the drone will make a controlled landing using its onboard sensors. Some of the cheaper drones may not have this feature, causing them to crash-land. If the adversary is well organized, he can place several sensors at some distance from each other, triangulating where the signal is coming from to track down the drone pilot. From this, conclusions can be drawn about the enemy's position, or a direct attempt can be made to take out the operator with grenade launchers or artillery.

High-powered drone jammers available on the market can operate over a distance of around 8 km and become increasingly effective the further the drone moves away from its pilot. The question remains as to how useful the jamming equipment ultimately is. There seems to be a gap between the extensive electronic warfare capability—certainly-of the Russians—via vehicle-based systems, which would be too expensive and at the same time too vulnerable targets for jamming small drones, and the increasingly numerous man-portable devices. The latter seem to be, according to public sources, almost "hobbyist" equipment.

Such simple jammers likely lack appropriate safety protocols, and Russian jammers are unlikely to have shielding and other design features that protect the operator's health. There is also always the question of the basic effectiveness of the device in question. Jammers may be an old technology—in use since World War II—but there are many modern aspects in terms of waveform and antenna design, jamming signal effectiveness, effective heat dissipation etc.

At the same time, you should never underestimate the enemy. Since the beginning of the Ukraine War, we saw in the Donbass that the Russians have managed to concentrate their electronic warfare (EW) and integrate it well into their ongoing operations. As a result, in some regions, every time Ukraine sends a small remote-controlled drone into the air, it is captured by Russian countermeasures and either immediately taken over or forced to return by cutting the electronic command link.

To avoid this, the small drones must not be remotely controlled, but programmed to fly a specific route so that only after they return can targets be identified when the footage is analyzed. This lengthy process is of limited value against moving targets. Delays in observation are also a major benefit to the adversary, who wants to protect his own troops. Which in turn could be undermined by escalation: Those who have loitering munitions available are also able to target and attack ECM systems.

Overall, small drones are still war consumables. In Ukraine, the assumed lifespan is one month.

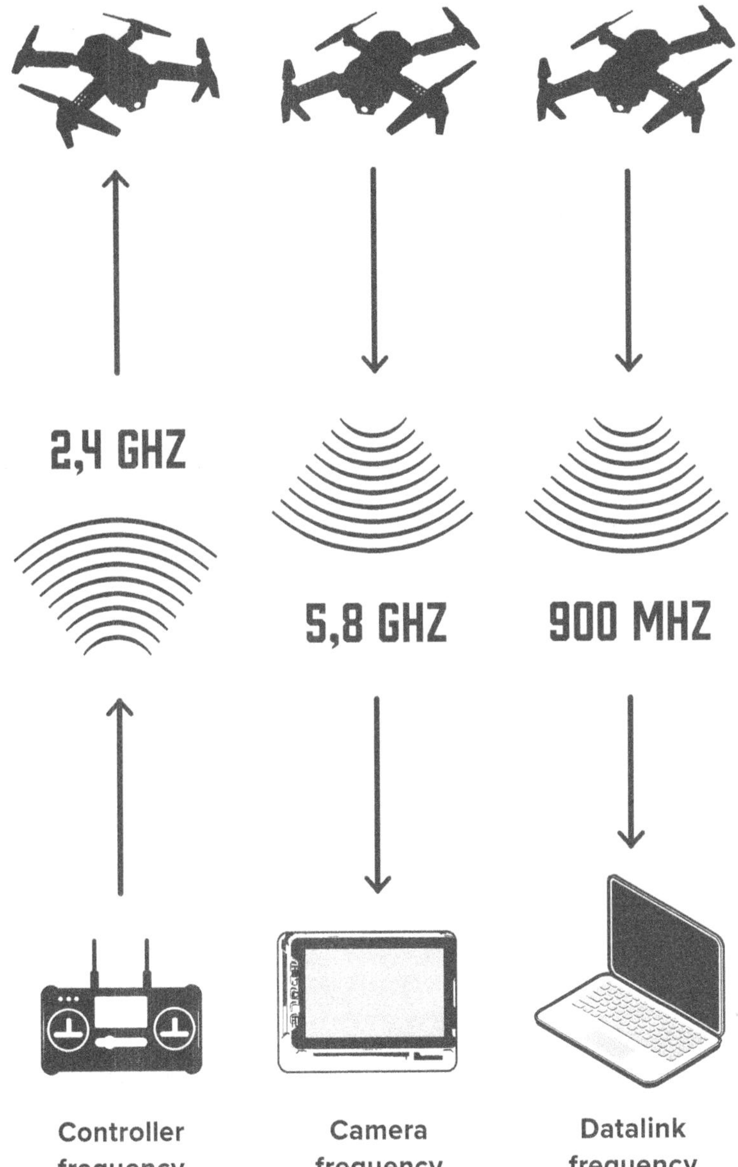
2,4 GHZ
5,8 GHZ
900 MHZ
Controller frequency
Camera frequency
Datalink frequency

10

NATURAL BORN KILLER: THE DRONE AS ATTACKER

IF YOU ARE ALREADY OVER THE ENEMY, IT IS EVEN QUICKER TO ENGAGE HIM WITH THE DRONE.

But while military drones are specially built or converted for this purpose, civilian aircraft are not designed for it. American "Predator" drones, controlled by pilots in containers, are not only professional flying machines with excellent optics and electronics, but can also be armed with missiles to liquidate the observed target. They have been used very frequently in the American "War on Terror." The "New York Times" had 1,300 documents from the operations in Syria, Iraq, and Afghanistan to evaluate: In five years, the U.S. Army flew more than 50,000 air strikes in these three countries. It was admitted that it has accidentally killed 1,417 civilians in airstrikes in Syria and Iraq since 2014. In Afghanistan, the official figure is 188 civilians killed since 2018, but the newspaper's research says the Pentagon's admitted numbers are "significantly understated." Since drones do not operate autonomously—the decision to attack is made by humans—there always remains the responsibility taken by pilots as well as politicians. Large drones are considered "clean weapons" that intervene efficiently. Weapon systems that can only be used by states are always image-makers, intended to avoid the use of soldiers on the ground. This is equally true of the victorious drones in the Georgia-Azerbaijan conflict (2021) and the Turkish "Bayraktar" drones on the side of Ukraine in the war with Russia (2022).

However, we deal here with small **UAS** of civilian origin. Most of the time, it is just about improving observation capabilities and how those can be exploited tactically. This is something that commercial drones can do perfectly well without modification, because it allows them to do exactly what they were made to do for civilians: Look at something from the air and take spectacular pictures and videos. At the same time, the drone is something that can be armed or become a direct attacker itself, as a kamikaze drone.

The **"Islamic State"** (ISIS) was the first terrorist organization to organize widespread use of drones. And it did so precisely because such devices are simple, cheap, and easy to replace. In the case of ISIS, it has also become clear that this was by no means a spontaneous improvisation: The terrorists obtained an improvised "air force," so to speak, via centrally organized supplies. The DJI "Phantom" was the dominant model on the market at the time, but not the only one. While the "Phantom" was modified off-the-shelf to drop small grenades, there was also larger equipment with the ability to drop an RPG warhead, which could certainly take out a tank from above. UAS were also widely used in the battle for Mosul. Drones, possibly quadcopters, reportedly dropped at least ten explosive devices during an hour-long engagement near the city's fourth Tigris bridge. (Incidentally, ISIS was not the first to use this tactic in Syria; Hezbollah made headlines the year before with a similar attack on rebels near Aleppo.) The highly organized drone air force of the "Islamic State" has yet to find imitators in this form, even in isolated instances among terrorist Islamists in Europe. Nevertheless, the main issue here is specialized technical knowledge rather than tactical experience; proliferation among potential users is not a major challenge. The basis of the weapon system, the commercial drone, is freely available. The rest is not difficult to implement for interested parties with an affinity for technology.

Currently, it is the war in Ukraine that shows small drones as weapons. Having already initially shown the value of commercial drones as observers and reconnaissance aircraft that are also efficient tools for guiding grenade launchers and guns, we are seeing more and more armed commercial drones as the front in the Donbass solidifies, with numerous variations and capabilities. While "Minis" are too small, "Mavics" and larger examples are quite veritable small-scale bombers. Ammunition variants that have been recognized are modified Western 40-mm grenades, Russian 25-mm VOG grenades (both with added stabilizing fins from the 3-D printer), and hand grenades that are dropped.

What is also bombastic is how eagerly the use of these weaponized flying devices is documented and spread via YouTube, TikTok, and other channels. This should alert security officials in Western Europe, because such explicit inspiration is bound to have repercussions. It is only a matter of time before someone uses the poor man's bombers for terrorist purposes (regardless of their ideological legitimacy).

For the conversion of drones and their armament, 3-D printers are the essential technology.

11 DRONE WARFARE: RECENT LESSONS FROM UKRAINE

QUANTITY IS A QUALITY OF ITS OWN, AND DRONES ARE WEAPONS.

Two years into the Ukraine War, we clearly see what roles drones can play and how important they are in this conflict. We are still talking about low-cost flying machines like the Chinese DJI "Mavic," with even high-end models costing only a few thousand dollars. In the field, they can provide rapid reconnaissance. This includes up-to-date situational awareness or warning of an assault or ambush, identifying targets for grenade launchers and artillery, producing quick videos for propaganda, or even the use as a flying explosive device guided onto the target in First Person View (FPV).

The number of drones needed is impressive: On the part of Ukraine, it is estimated that up to 10,000 drones are "consumed" per month—that would be more than 330 per day. If we take into account that the Russians have strongly followed suit with drones, we may approximately double the figure. The logistics are exciting as well: On both sides, it is often crowdfunding that guarantees drone supplies. At the same time, the mass of commercial, basic equipment comes from China—for both sides, which also says a lot about the conflict.

However, information collected by drones is only useful if it is shared effectively with combat units. That's where Starlink, Elon Musk's SpaceX satellite network, comes in. With more than 4,000 near-Earth satellites providing off-grid Internet access, Starlink is a lifeline for data. Starlink facilitates the rapid exchange of drone imagery and relays targets to the nearest artillery team, revolutionizing battlefield communications.

Now that both sides know how dangerous drones are, defenses have evolved as well. Drones are shot down from the ground or brought down by electronic warfare, disrupting their GPS navigation, or cutting off radio communications with the operator. When contact with the drone pilot is lost, the craft smashes into trees or simply hovers until its batteries run out, and it falls to the ground. In places like Bakhmut, where some of the fiercest fighting of the war took place, Russian interference was so intense that DJI "Mavics" and the like could only move a few hundred meters away from their operators without losing connection.

"In December, we could still fly three kilometers," a Ukrainian drone operator told a British newspaper in April 2023. "Now the guys say they can't fly further than 500 meters."

So the next logical step in the mini-drone arms race might be to harden drones against electronic warfare attacks, but experts say that's not worth it. The pragmatic solution: Buy more and fly more. In war, quantity is a quality of its own. The cheaper and more numerous, the more missions can be flown. The loss of one or thousands of cheap systems does not hurt as much as the loss of a more expensive one. The cheapest "Bayraktar" costs at least a million dollars, but you can get more than 500 "Mavic Pros" it.

At the same time, you can see a dramatic increase in the use of commercial quadcopters and FPV drones as flying weapons in Ukraine. Again, quantity that is the winning factor, along with targeting accuracy. Both sides have already deployed thousands of civilian drones with improvised warheads. Colloquially, they are called "kamikaze drones" because the warhead destroys the drone itself. They are FPV drones, because the operator wears goggles that show a video feed during flight. He flies with a view from the "cockpit." The operator can carry out attacks this way. In one clip posted on the Internet, FPV drones dive into a tunnel to attack a Russian tank. In others, they race after vehicles and destroy them. Faster, more maneuverable, and cheaper than other drones, they could change warfare, analysts say. "Poor man's loitering munitions," one might say. Both armies use quadcopters that drop grenades. However, FPV drones require an experienced remote pilot and are designed for speed and maneuverability. The Ukrainians have home-built versions that are now being copied by the Russians. They are assembled by volunteers or by the soldiers themselves from components provided by donors. The simplicity of the electronics and the use of commercially available parts make production inexpensive. A "Pegasus" attack drone manufactured in Ukraine comes in at $462. The larger and more sophisticated "Switchblade" drones America supplies to Ukraine, which carry only a small antipersonnel warhead, is $52,000. The low cost of FPV offsets its relatively low success rate in destroying targets. Operators estimate the FPV success rate at 50–80 percent, compared with over 90 percent for U.S. "Javelin" anti-tank missiles.

A key lesson remains, however, that the improvised mass use of civilian drones may well upstage "gold-plated" procurement programs of the kind a lot of Western countries try to develop and field.

FURTHER READING

DRONES ARE A MODERN TOPIC, YOU CAN FIND LOTS OF INFO AND MATERIAL ON YOUTUBE AND SOCIAL MEDIA.

Nevertheless, books are a good source for acquiring compact knowledge. We have divided this section into two categories, namely "tactical" (i.e., application-oriented, civilian introductions) and "strategic" (literature that seeks to classify the topic of drones on the battlefield as a whole).

+ THE DRONE PILOT'S HANDBOOK

A practical paperback book that introduces you to flying drones. Setup is a bit like a children's book, with quite a few illustrations, but in between is concise, sensible text that quickly explains to you the basics of civilian drone flying: From the different types, to movements in 2-D and then 3-D, to using the drone as a camera, and of course everything about the legalities. Not much to read, but lots of useful knowledge for beginners.

"The Drone Pilot's Handbook. Basics, Practice, Rules, Technology" by Adam Juniper and Rena Herbig. Bassermann Verlag, Munich 2019, 160 pages, €10.27

+ DRONES FOR DUMMIES

Ready to step up to the world of unmanned aerial vehicles? "Drones for Dummies" introduces you to the fascinating world of UAS in the style of the "Dummie" series, so it's all down to earth. The book is written in plain English and provides comprehensive coverage of all the basics.

*"Drones For Dummies"
John Wiley & Sons, Hoboken 2015, 286 pages, €18.85*

+ DRONE LICENSE COMPACT

Everything for anyone who wants to fly the biggies in a civilian capacity: The new EU Remote Pilot License has been mandatory for drones weighing more than

250 g since 2021. All content is based on the examination requirements for the EU remote pilot licenses A1, A3, and A2. In addition, relevant factors for the commercial operation of drones are considered, e.g., copyright and personal rights in aerial photography.

"Drone Pilot License Compact. The textbook on drone flight" by Andreas Platis and Uwe Nortmann. Motorbuch Verlag, Stuttgart 2021, 240 pages, €30.73

+ POSSIBILITIES AND LEGAL FRAMEWORK FOR THE USE OF DRONES BY AUTHORITIES AND ORGANIZATIONS WITH SECURITY TASKS

The drone book with the most beautiful title of all. This book is a guide for authorities and organizations with security tasks and also provides an outlook on future supranational regulations for the operation of UAS. On the one hand, this guide analyzes the scope of special rights, on the other hand it offers more legal certainty in the use of UAS. In addition, it shows the potential the use of UAS can provide for civil protection.

"Possibilities and legal framework for the use of drones by authorities and organizations with security tasks" by Maximilian Beck. BoD, Norderstedt 2020, 308 pages, €30.73

MORE KNOWLEDGE

In addition to books, current magazine articles provide knowledge on the use of drones. There are plenty of professional journals that cover the topic. We have collected essential articles (URL shortened):

"Micro-UAV Augmented 3D Tactics"
in Small Wars Journal: ly/3KpjzhO

"Drone and Counter-Drone Warfare At Tactical Level"
in Land Warfare Studies: bit.ly/3KobA3T

"How will drones affect infantry tactics?"
in Defence iQ: bit.ly/3QWXF7n

"On Drones and Tactics"
at Modern War Institute: bit.ly/3AShRBy

"Lessons from Use of Drones in the Ukraine War"
at Droneshield: bit.ly/3ARqOuT

"How DAESH uses Drones in the Battlefield"
in Politics Today: bit.ly/3RhROo8

"Why Drones Have Not Revolutionized War"
at MIT: bit.ly/3RjpK8O

Russians With Attitude @RWApodcast · 1. Mai

Most important military lesson so far seems to be that recon needs to be as decentralized as possible & that every battalion or at least brigade level unit needs a dedicated aerial recon platoon with fancy drones & every platoon needs a drone operator with a smaller/cheaper drone

59 188 1.611

Russians With Attitude @RWApodcast · 1. Mai

The DJI Mini 2 has shown itself to be an incredible asset on the modern battlefield. Both sides use it extensively to support all kinds of tactical operations; it is equally important for both infantry & armor & turns a 60 years old Soviet howitzer into a 152mm sniper rifle.

33 194 1.200

13

MARKUS REISNER:

USE OF DRONES IN THE UKRAINE WAR

Observations of the war in Ukraine often focus on land forces, tanks, and ballistic missiles. However, a comprehensive analysis must also consider the use of drones. They have become an indispensable part of modern warfare.

The term "drone" covers a wide spectrum of unmanned aerial objects. This ranges from small, commercially available drones (e.g., Chinese DJI drones) to military drones the size of commercial aircraft. The classification can be based on different aspects, such as:

+ **WEIGHT,**
+ **PERFORMANCE,**
+ **FLIGHT DISTANCE,**
+ **SIZE OR**
+ **TYPE (FIXED-WING OR ROTARY-WING).**

CLASSIFICATION OF DRONES BY WEIGHT

Class 1:	under 150 kg	micro, mini, small
Class 2:	150 to 600 kg	tactical drones with longer range
Class 3:	over 600 kg	MALE (Medium-Altitude Long Endurance, comparable to "Predator" or "Reaper" of the U.S. Armed Forces) with the possibility of carrying air-to-ground missiles, HALE, STRIKE, or COMBAT.

The RQ-4B "Global Hawk" drone, which is the size of a commercial airliner, for example, is categorized as Class 3 and is used by the U.S. and NATO. It also plays a role in the Ukraine war, being used over the Black Sea to reconnoiter Russian forces in southern Ukraine.

The RQ-4B Global Hawk is an unmanned long-range reconnaissance aircraft that can be deployed autonomously and with satellite-feed on operations.
(Photo: U.S. Air Force/Bobbi Zapka; public domain)

The smaller and cheaper types require a lower technical effort. The larger, the greater the effort and price, but the benefit is a wider range of possible uses. Depending on the type of mission, it is also decided whether to use a reconnaissance drone or an armed drone. Pure reconnaissance drones can be procured by the Ukrainian armed forces themselves (through crowdfunding, for example). Larger military drones, which can also be armed, are brought into the country primarily through arms shipments from other nations.

WHAT DRONES DOES RUSSIA USE?

The Russian armed forces have various Class 1, 2, and 3 drones, including tactical and smaller drones (e.g., "Forpost" or "Orlan-10") but also larger models (e.g., "Orion," comparable to "Predator" of the U.S. armed forces). These drones have already been used in Syria and, since 2014, in Ukraine. For example, Russian drones (e.g., "Forpost") reconnoitered Ukrainian forces near the border as early as 2014, which were then targeted through the use of artillery and multiple rocket launchers.

Russia also deploys loitering munitions or loitering weapons, also known as kamikaze drones. These are guided weapons that are launched without a specific target and then circle ("loiter") over an area until a target is acquired. When they locate a target, they strike. In the Ukraine war, for example, Russia deployed the "KUB-BLA" model in Kyiv. Drone reconnaissance combined with artillery fire proved effective for Russia. However, electronic warfare (EW) with the support of drones can also look different. For example, as part of an EW complex, the "Orlan.

WHAT DRONES DOES UKRAINE USE?

Ukraine has been able to learn from the use of drones against its armed forces and improve its capabilities in terms of resilience and the use of drones. It has some Class 1 and Class 2 drone models but has also received donations or deliveries of other systems (e.g., "Quantix Recon" or "Vector") since the start of the war. Ukraine, for example, procured the Turkish model "Bayraktar TB2." It is characterized by its small, compact design and the ability to carry air-to-surface missiles. "Bayraktar TB2" can deploy MAM-L, MAM-C and MAM-T ("Mini Akıllı Mühimmat"—"smart micro munitions" for drones) missiles. In 2020, Turkish drones were used for the first time when Ukraine used them to observe radar stations in Crimea. A year later, the "Bayraktar TB2" was used to strafe an artillery position in the separatist areas of eastern Ukraine. This drone has also been engaging targets on the Russian side since the war began on February 24, 2022.

Ukraine also has loitering munitions, like the Polish "Warmate" model. This drone plunges itself into a target and destroys it on impact. The warhead even allows it to penetrate strong armor. The delivery of NATO and U.S. models is expected. Specifically, these are the "Switchblade" 300 or 600 systems and the "Phoenix Ghost." Both systems could have a major effect on the battlefield.

TACTICS: POSSIBILITIES OF DRONE USE

There are many ways to use drones. The following ten tactics are commonly employed:

- + RECCE/SUPPORT
- + TARGETED STRIKE
- + AREA STRIKE
- + STRIKE [UNGUIDED]
- + STRIKE [GUIDED]
- + STRIKE [LOITERING]
- + STRIKE [LONG RANGE]
- + STRIKE [HIGH-VALUE TARGETS]
- + ELECTRONIC WARFARE
- + SWARM

+ RECCE/SUPPORT

Classical reconnaissance.

Example: ZALA drones support Russian forces by providing improved situational awareness.

+ TARGETED STRIKE

Targeted use of artillery based on the acquired data.

Example: The drone reconnoiters a target, and the artillery shell strikes there in a controlled manner. The drone may even designate the target. A whole range of models are suitable for this, such as the commercially available "Mavic Pro" system.

+ AREA STRIKE

Use of the drone for reconnaissance for artillery systems that achieve an area effect.

Example: The Russian "Forpost" drone generates data on the basis of which an artillery battery (e.g., 2S19 guns) can subsequently engage a target.

+ STRIKE [UNGUIDED]

The drone is armed with varying ordnance.

Example: Ukraine used multicopters loaded with droppers in the first weeks of the war. With appropriate penetration, it was even possible to destroy mechanized targets.

+ STRIKE [GUIDED]

Surface-to-surface drones with payloads are used to fly targeted attacks.

Example: The UJ-22 drone used by Ukraine.

+ STRIKE [LOITERING]

Drone that launches itself into the target and destroys it on impact.

Example: Loitering munitions developed by Ukraine (e.g., ST-35) have a penetrating power that can destroy even strong armor.

+ STRIKE [LONG RANGE]

Drone systems that carry air-to-surface missiles can engage targets over a long distance.

Example: The "Bayraktar-TB2" drone, which is difficult to detect because it is small and compact and has a small radar echo. As a result, the missile's impact comes as a surprise because the enemy cannot locate it.

+ STRIKE [HIGH-VALUE TARGETS]

The drone is used to reconnoiter a specific target for an extended period of time and then strike with precision.

Example: The Russian "Orion" drone is suitable for attacking a high-value target with its air-to-ground missile. There was circumstantial evidence of this during combat in the Kyiv area. Here, Ukrainian President Volodymyr Zelenskyy was seen appearing in front of a blue screen several times. This indicates that he did not go out into the open to avoid being a target for Russian drones.

+ ELECTRONIC WARFARE

Drone as a platform or part of electronic warfare.
Example: The Russian drone "Orlan-10" acts like a flying cell phone tower as part of the EW complex RB-341. Cell phones can force-dial into the drone and use it to send targeted messages accordingly.

+ SWARM

Deployment of multiple drones in a swarm. When the drones communicate with each other, it is called an intelligent swarm. When multiple drones are deployed simultaneously, it is mainly the quantity that makes it possible to reach the target.
Example: It is possible that the Ukrainian armed forces will deploy the "Switchblade" or "Phoenix Ghost" drones in large numbers to overwhelm enemy defense systems and gain an advantage on the battlefield. Thus, the use of drones in a swarm could be a tactical "game changer."

"GAME CHANGER" IN WARFARE?

The use of kamikaze (loitering munitions) drones in swarms begs the question whether the tank will continue to have any relevance on the battlefield in the future.

The reality is that tanks remain the only way to take possession of terrain with a combination of armor, firepower, and mobility. To be able to do that in the future, mechanized formations, for example, will need to be appropriately protected against drone threats on their march to the objective. One of the challenges is the very small radar echo of kamikaze drones. All defense systems riding along must nevertheless be able to detect and counter drones. If many drones are deployed simultaneously, the defense systems must be designed to counter them. In this way, the tank can retain its importance in the future. An example of this is the Russian "Tor-M1" system, which is also being used in Ukraine. This was retrofitted based on experiences in Syria to be used specifically against small drones.

Another factor is the complexity of drones. The Russian "Orlan-10", for example, has very simply designed components, most of which are civilian rather than military. This has advantages and disadvantages. For example, they may not reach the quality of military components, but they can be easily manufactured with commercially available products (e.g.,

Chinese engines, components from model aircraft) and adapted into a powerful military drone. Such systems can hardly be "countered" by sanctions because the components are available on the civilian market.

CONCLUSION:

The Ukraine war example demonstrates the transformation of modern warfare. Drones can provide a decisive advantage on the battlefield in both reconnaissance and attack. At the same time, however, challenges arise in terms of procurement, operation, and deployment. Moreover, an ethical and moral dilemma opens up if such systems in future versions also perform actions autonomously—without human intervention. International humanitarian law, for example, does not yet explicitly address such weapons.

Col dG Dr. Markus Reisner, PhD, head of the development department at the Theresian Military Academy.

First published on TRUPPENDIENST:

truppendienst.com

Reprinted with kind permission.

LESSONS LEARNED FROM UKRAINE

U.S. soldiers and others observing the war in Ukraine can take away some important lessons from the conflict so far regarding UAS on the battlefield, *Task & Purpose* reports.

ALWAYS ASSUME THAT SOMEONE OR SOMETHING IS LOOKING FOR YOU.

Be it visually or with direct-finding, communication, and signal countermeasures. Proximity to the forward line of troops means the enemy is trying to destroy high-gain targets, such as command and control points, communications facilities, and fire support units. Camouflage is a must, and limiting ground movement during the day is also essential. Moving units under cover of darkness (again, UAS night vision capability is not assumed) also reduces the signature.

CREATE A PLAN TO DEFEND AGAINST UAS.

Although some platforms can fly higher than conventional weapons can reach, a small COTS drone can be shot down with small arms. Knowing that these platforms are available to even the smallest unit and that their reconnaissance capabilities (even if it's just video taken with an iPhone) provide real-time coverage means that a friendly unit can be tracked down and engaged. Early warning systems and radar to detect these drones will also be critical. These must be easily transportable and deployable on short notice. Treat any aerial observation of your unit and its maneuvers as suspicious.

SEEK YOUR OWN UAS ADVANTAGE.

Do that either through a unit-supported collection plan or procure a drone, if possible, to support your own maneuvers and command and control. This can and should work both ways, although U.S. soldiers will likely have to endure cumbersome approval processes for allocations, and homemade arming solutions for unit-level COTS UAS will undoubtedly be frowned upon. However, necessity is the mother of invention. I would explore all options to provide for my unit's defense and protection. Commanders have all sorts of creative people working for them, and I recommend talking to your own people to see what experience they can bring to the table in providing solutions. At the sUAS level, neighboring unit coordination and disentanglement are extremely important.

▶

UNDERSTAND YOUR PART OF THE BATTLEFIELD.

Leaders should always provide intelligence before acting. At the smaller unit level, they should try to learn as much as possible about the area in which they are fighting, including enemy movements and defensive positions. With the number and type of platforms available, some type of observation system is available at the operational level. At the tactical level, which is perhaps most significant for UAS operations, leaders gain knowledge of their immediate operational space with their own integrated sUAS and have an up-to-date situational picture that expands their decision-making latitude, often before their higher commander understands.

From the article "Off the shelf, above the fight. How cheap drones are completely changing warfare;" taskandpurpose.com/opinion/drones-uas-warfare-ukraine-russia

APPENDIX

- THE THREAT SURVIVAL GUIDE
- THE AL-QAEDA PAPERS: DRONES
- DJI STATEMENT ON MILITARY USE OF DRONES

FOR UNOFFICAL USE ONLY (FUUO)

X47C
Military Surveillance / Attack
US

Meter
0 2 4 6 8 10

Feet
0 5 10 15 20 25 30 35

Sentinel
Military Surveillance
US

Global Hawk
Military Surveillance
US/KR

Giant Eagle
Military Surveillance
CN

Pterodactyl I
Military Surveillance / Attack
CN

Predator
Military Surveillance / Attack
US/IT/MA/EG/SA/AE/PH

Fire Scout
Military Surveillance / Attack
US

Shadow
Military Surveillance
US/AU/IT/PK/RO/SE

Rustom I
Military Surveillance
IN

WASP III
Military Reconnaissance
US/FR/AU/SE

Harpy
Military Attack
IL/TR/CN

Scan Eagle
Military Surveillance
US/GB/CA/MY/CO/NL/JP/[illegible]/SG/
TN/AU/IT/RO/ES/YE

Killer Bee
Surveillance
US/TR

Raven
Military Reconnaissance
+20 countries

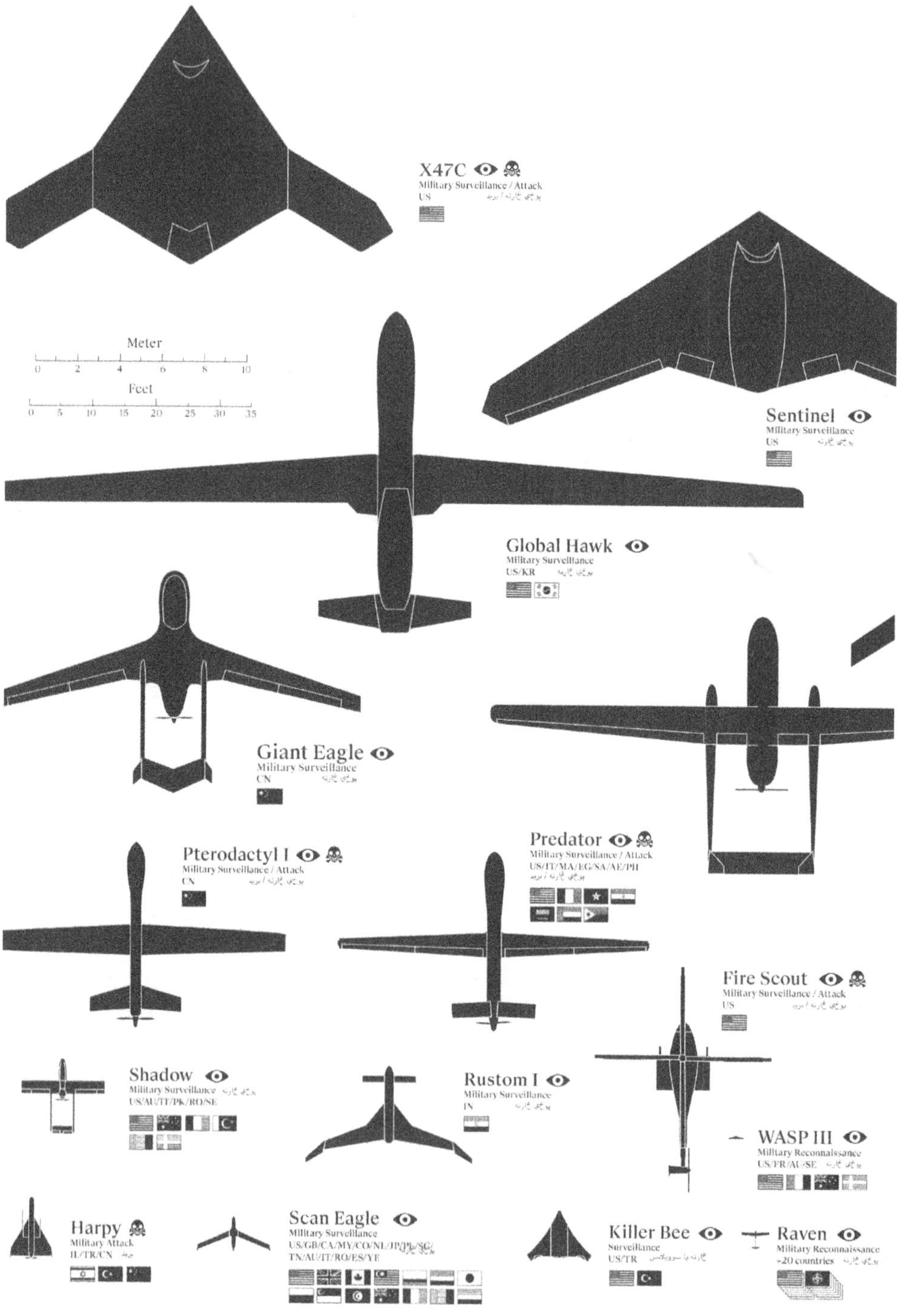

د بې پیلوټه الوتکو د پایښت لاربنود

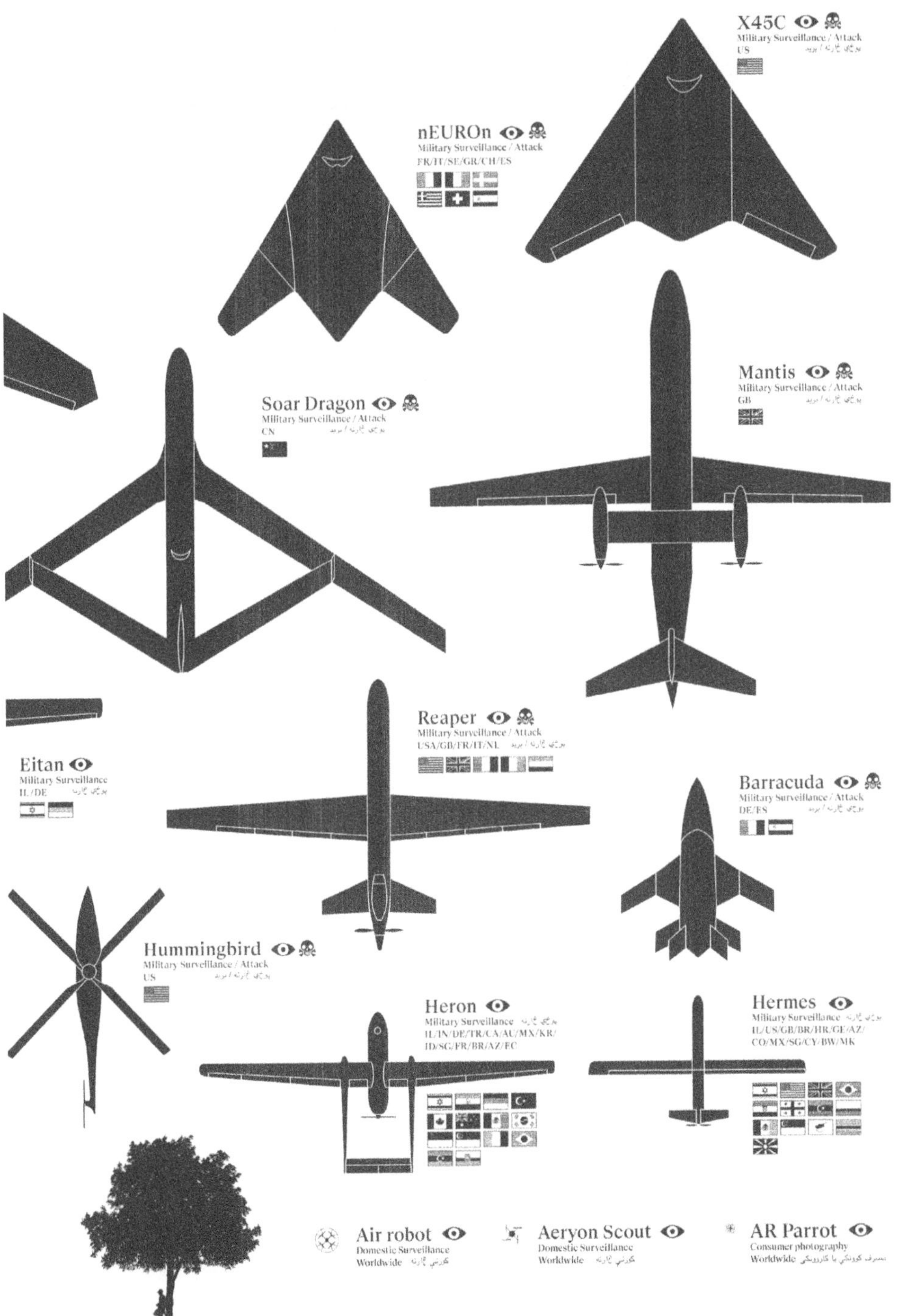

THE THREAT SURVIVAL GUIDE

BIRD WATCHING IN THE 21ST CENTURY

Our ancestors could recognize birds of prey from a distance by their silhouette. But do we recognize modern predators similarly well today?

Drones are remotely piloted aircraft used for surveillance, killing, rescue missions, and scientific research. Most drones are used by the armed forces. And there are more and more of them.

In 2012, the Federal Aviation Administration (FAA) predicted that as many as 30,000 drones could be in the air in 20 years—over U.S. territory alone. So, with robotic birds soon to become commonplace, we should be prepared to be able to identify them. This survival guide is an attempt to familiarize ourselves and future generations with a changing technological environment.

This document includes silhouettes of the drone species that are currently most common and will be in the near future. Each illustration also provides information on which nation is using the drone and whether it is for surveillance or killing. All drones are shown at scale to allow conclusions about their sizes. From the smallest consumer drone, less than one meter long, to the "Global Hawk," with a wingspan of nearly 40 meters. To make this document available to a wide readership, it can be downloaded as a PDF or DOC. Additional translations will be available over time.

DISCLAIMER:

This document is intended for informational purposes only. We do not condone or support any form of violence, neither against people nor property. All content may be freely shared, edited, and translated for non-commercial use and is available free of charge on the dronesurvivalguide.org website.

HOW TO HIDE FROM DRONES

Drones are equipped with high-powered cameras that can detect people and vehicles from heights of several kilometers. Most drones have night vision and/or infrared cameras, called **FLIR sensors**. These can spot human heat signatures from great distances during the day and at night.

STILL, THERE ARE WAYS TO HIDE FROM DRONES:

---

1. **Daytime camouflage: Hide in the shade of buildings and trees. Use dense forests as natural camouflage, or use camouflage nets.**

---

2. **Camouflage at night: Hide in buildings or under the cover of trees and bushes. Do not use flashlights and keep car headlights off, even on long trips. Drones on night missions can easily locate these light sources.**

---

3. **Thermal camouflage: MPET space blankets block infrared radiation. When you use a space blanket as a poncho at night, you shield your heat signature from infrared detection. In the summer, when the temperature is between 36 °C and 40 °C, IR cameras cannot distinguish between body heat and ambient heat.**

---

4. Wait for bad weather. Drones cannot be used in high winds, smoke, heavy rain, or severe weather in general.

5. No wireless communication. The use of cell phones or GPS-based communication will give away your location.

6. Reflective material placed on cars and rooftops can mislead a drone's camera.

7. Dummies: Use mannequins or life-size decoys to fool the drone.

HOW TO HACK DRONES

Drones are controlled remotely. The pilots who control the drones may be located at ground stations thousands of miles away. The control link is the satellite-established data connection that the pilot uses to control the aircraft. By disrupting or intercepting this data transmission, the drone's control can also be disrupted. While the data link can be encrypted, it often is not.

The drone sees, even when it is not seen: This is an observation view through thermal imaging optics.

1. **Interception: One sophisticated method is to intercept drone frequencies using sky-grabbing software, a satellite dish, and a TV tuner. Both incoming and outgoing drone communications can be intercepted this way.**

2. **Jamming: By transmitting on different frequencies or an entire frequency range, the radio link between the drone pilot and the drone can be cut.**

3. **GPS spoofing: Small, portable GPS transmitters can be used to send false GPS signals that interfere with the drone's navigation system. In this way, the drone can be made to crash, for example, or even hijacked and landed on a runway.**

Sources:
Mike Adams, "Health Ranger's Intelligence Analysis of Military Drones: Payloads, Countermeasures, and More," naturalnews.com, July 16, 2012.
"The Al-Qaeda Papers—Drones," Associated Press, February 2013 (this is a document discovered by the Associated Press in buildings in Timbuktu, Mali, which had been occupied by al-Qaeda fighters shortly before).
"Evading Thermal Imaging and Radar Detection," United States Militia, Special Forces.

THE AL-QAEDA PAPERS: DRONES

(As published by AP)

This document is one of several found by "The Associated Press" in buildings recently occupied by al-Qaida fighters in Timbuktu, Mali. Below is a translation in English.

In support of Ibyan province (Yemen) Military Research Workshop.

I have said in my article "Strategies of Capabilities for Ansar al-Sharia" that the American retaliation against the Mujahideen military movements in Ibyan province will be restricted to the war of the drone. My expectations have been assured after the recent "New York Times" leakage that the CIA will handle the situation, and for this, it set up a secret military base for the drones in a neighboring country.* It is important now that we understand this American army strategy and discuss ways to disable this strategy.

To start with, we have to know that the Americans did not resort to this approach—the war of the drone—because they have shortages in the combat jets like the F-16 and other types or they don't possess enough troops, but because it is the most suitable approach for them now. The Americans fully realize that they are in the 10th year of war and that they were economically exhausted and suffered human losses and they were confronted with public pressure backed by the Congress in a way that it made the honorable and responsible withdrawal from the war as a prime goal of the White House. But this does not mean that they are abandoning the war, rather, they pushed them to seek alternative military strategies that enable them to continue the war without being economically depleted or suffer human losses and avoid the American public opinion pressure. Here the war of the drone appeared as a perfect solution. The drone is unmanned and costs nothing compared to manned jets, and it does not create public exasperation when it crashes because the increase of human losses in the past pushed the American people to take to the streets shouting "Bring back our sons," and if a drone crashes, no one will shout "Bring back our planes!"

In comparison, the cost of 1,000 drones equals the price of an F-15 "Eagle" jet. If we talk about the latest models, like the "Predator," it costs $10 million while the cost of an F-16 is $350 million and the fuel for 200 flights of a drone equals the fuel consumed by one flight of an F-4 "Phantom" jet. The training of a pilot of a "Tornado" costs 1 million Pound Sterling while training a drone operator costs nothing and it takes only three months. Therefore, the Americans have chosen a comfortable war to prove to us their indifference to a long war. For this, they appointed the commander of American forces in Afghanistan [David] Petraeus as CIA director to lead the war from there and they have already tried this strategy in Waziristan that proved successful, and they are going to apply it now in Yemen. So what are they going to do? I believe that foiling this strategy depends on three things: The formation of a public opinion to stand against the attacks, deterring of spies, and tactics of deception and blurring.

These tactics are:

1 – It is possible to know the intention and the mission of the drone by using the Russian-made "Sky Grabber" device to infiltrate the drone's waves and frequencies. The device is available on the market for $2,595 and the one who operates it should be a computer-know-how.

---

2 – Using devices that broadcast frequencies or packs of frequencies to disconnect the contacts and confuse the frequencies used to control the drone. The Mujahideen have had successful experiments using the Russian-made "Racal."

---

3 – Spreading the reflective pieces of glass on a car or on the roof of the building.

---

4 — Placing a group of skilled snipers to hunt the drones, especially the reconnaissance ones because they fly low, about six kilometers or less.

---

5 – Jamming of and confusing of electronic communication using the ordinary water-lifting dynamo fitted with a 30-meter copper pole.

---

6 – Jamming of and confusing of electronic communication using old equipment and keeping it running 24 hours because of their strong frequencies. It is also possible using simple ideas of deception equipment to attract the electronic waves, devices similar to that used by the Yugoslav army when they used the microwave [oven] to attract and confuse the NATO missiles fitted with electromagnetic searching devices.

---

7 – Using general confusion methods and not use permanent headquarters.

---

8 – Discovering the presence of a drone through well-placed reconnaissance networks and warning all the formations to halt any movement in the area.

---

9 – To hide from being directly or indirectly spotted, especially at night.

---

10 – To hide under thick trees because they are the best cover against aircraft.

---

11 – To stay in places unlit by the sun, such as the shadows of buildings or trees.

---

12 – Maintain complete silence of all wireless contacts.

---

13 – Disembark vehicles and keep away from them, especially when being chased or during combat.

---

14 – To deceive the drone by entering places with multiple entrances and exits.

---

15 – Using underground shelters, because the missiles fired by these aircraft are usually fragmentation anti-personnel projectiles and not anti-buildings.

---

16 – To avoid gathering in open areas and in urgent cases use buildings with multiple doors or exits.

---

17 – Forming anti-spies groups to look for spies and agents.

---

18 – Formation of fake gatherings such as using dolls and statues to be placed outside false ditches to mislead the enemy.

---

19 – When discovering that a drone is after a car, leave the car immediately and everyone should go in a different direction, because the planes are unable to get after everyone.

---

20 – Using natural barricades like forests and caves when there is an urgent need for training or gathering.

---

21 – In frequently targeted areas, use smoke as cover by burning tires.

---

22 – As for the leaders or those sought after, they should not use communications equipment because the enemy usually keeps a voice tag through which they can identify the speaking person and then locate him.

Deterring the spies

The drones used in the attacks in Swat Valley depend on electronic chips or radioactive dyes placed at the target by the spy or the agent, then the guided missiles come directly toward these targets. The spy, therefore, is the main pillar of this operation, so there is need to resort to decisive deterrents against anyone who might dare to carry out such a mission, he must be hanged in public places with a sign hanging from his neck identifying him as an "American spy," or any other deterrent similar to that done to [Israeli spy hanged in Syria] Levy Cohen or [late Afghan President] Najibullah.

The formation of public opinion against these attacks to instigate an alternative Arab and Islamic street.

I think these measures are capable, with God's help, of disabling the new strategy of the American army in the medium or long-range levels. This is not all we have. There is the golden solution that shortens the long distances and through which we can bring back the pressure of the American public opinion in a more active way depending on the strategy of kidnapping in exchange for the drone strategy, and we should not stop until they stop their strategy which will enable all the supporters of jihad to take part in defeating Petraeus and his new strategy. We start kidnapping Western citizens in any spot in the world, whether in the Islamic Maghreb, Egypt, Iraq, or other easy kidnapping places, and the only demand is the halt of attacks on civilians in Yemen, which is a just and humanitarian demand that will create world support and a public opinion pressure in America as they are being hurt again. We, therefore, aim at the core of the nation's strategy. If it fails, America will accordingly collapse. We are also taking part in laying a block in the promising Islamic State in the Arab peninsula.

In case there are any other tactics or deterring means, please add them here so that the benefit will be wider, and I pray to God to save us from the American intrigues and turn these intrigues against them.

Written by Abdullah bin Mohammed

Date 15 Rajab, 1432 (Islamic calendar), corresponding to June 17, 2011.

* *The "Associated Press," the "New York Times," and other media reported the construction of a drone base to target Yemen in June 2011, but withheld the exact location at the request of senior administration officials. The "New York Times" disclosed the location earlier this month as Saudi Arabia.*

dji

16 DJI STATEMENT ON MILITARY USE OF DRONES

More than 15 years ago, DJI was founded to explore the astonishing new possibilities of drone technology. From our first attempts at building tiny helicopters to our latest cutting-edge drones for professionals, we have been driven not only by the technical challenge of robotic innovation, but by how this new technology can help people. And with each improvement and advance, our drones have helped make the world better. Our products have enabled new ways to see our world, a new storytelling language for creators, new business and job opportunities, new techniques for growing food and protecting our environment, and new capabilities for responding to emergencies. Our drones have preserved endangered species, saved global landmarks from destruction, and saved lives all over the world.

With that in mind, we want to reiterate a position we have long held: Our products are made to improve people's lives and benefit the world, and we absolutely deplore any use of our products to cause harm. DJI has only ever made products for civilian use; they are not designed for military applications.

Specifically:

- **DJI does not market or sell our products for military use.**
- **DJI does not provide after-sales services for products that have been identified as being used for military purposes.**
- **DJI has unequivocally opposed attempts to attach weapons to our products.**
- **DJI has refused to customize or enable modifications that would enable our products for military use.**

DJI believes strongly in these principles. Our distributors, resellers, and other business partners have committed to following it when they sell and use our products. They agree not to sell DJI products to customers who clearly plan to use them for military purposes, or help modify our products for military use, and they understand we will terminate our business relationship with them if they cannot adhere to this commitment.

DJI is dedicated to creating products that benefit society. We take great pride that our drones have rescued people who were lost and near death, enabled scientists to protect our environment, enabled enterprises to improve workflows and reduce risk in challenging times, and created enjoyment for millions of people around the world. We will never accept any use of our products to cause harm, and we will continue striving to improve the world with our work.

April 21, 2022
dji.com

THE AUTHORS

CHRISTIAN VÄTH

has been a freelance head instructor since 2017, educating civilians and LE/MIL units on infantry-related topics like marksmanship, small unit tactics, force-on-force, TCCC, and combat mindset.

Former active-duty infantry officer (2008–2021), most recently responsible for training at the premier marksmanship unit at the German Federal Infantry School.

Command positions in infantry units of the German Army, the British Royal Marines, and the Norwegian Telemark Battalion since 2008. Staff officer of the Army Reserve since 2021.

Väth holds a master's degree in history (Helmut Schmidt University of the German Federal Armed Forces and University of Liverpool).

lightinfantry.de

MARKUS REISNER

Colonel in the General Staff of the Austrian Armed Forces. PhD in both history and law from the University of Vienna. Vast experience from numerous deployments in Bosnia and Herzegovina, Kosovo, Afghanistan, Iraq, Chad, Central Africa, and Mali. Research interests: deployment and future of unmanned reconnaissance and weapon systems; historical and current military issues.

Author of several books. Head of the Research Department of the Theresian Military Academy in Wiener Neustadt since June 2020 and CO of the Austrian Honor Guard since 2022.

Printed in the USA
CPSIA information can be obtained
at www.ICGtesting.com
CBHW081057041024
15325CB00036BA/191